INTERCROPPING AND THE SCIENTIFIC BASIS
OF TRADITIONAL AGRICULTURE

Intercropping and the Scientific Basis of Traditional Agriculture

DONALD Q. INNIS

INTERMEDIATE TECHNOLOGY PUBLICATIONS 1997

Intermediate Technology Publications Ltd,
103–105 Southampton Row, London WC1B 4HH, UK

© Wendy H. Innis, 1997

A CIP record for this book is available from the British Library

ISBN 1 85339 328 2

Typeset by J&L Composition Ltd, Filey, N. Yorkshire
Printed in UK by SRP, Exeter

Contents

Introduction		vi
1	The nature of intercropping	1
2	Intercropping in the Christiana area of Jamaica	34
3	Intercropping in Nepal	66
4	Intercropping in India	76
5	The Earth's soil and its future	102
6	Social and economic implications of agricultural practices	107
Appendix 1, Land equivalent ratios		113
Appendix 2, Plant names		163
Bibliography		165

Introduction

THIS VOLUME ON 'Intercropping and the Scientific Basis of Traditional Agriculture' in the new *IT Studies in Indigenous Knowledge and Development* encompasses a true pioneering, comparative study of the practice of growing two or more crops at once in the same field, which was conducted in many areas of the developing world by the late Donald Innis, who died shortly after finishing the manuscript.

By the time that the attention to several forms of indigenous knowledge systems was gaining increased recognition in sectors of the international development arena such as human and animal health, agriculture, forestry, fisheries and natural resources management, Donald Innis had already undertaken long-term empirical studies of how small farms managed to maintain and conserve the use of soil without non-renewable resources in many countries around the globe including Jamaica, Nepal and India. Before the Brundtland Report on '*Our Common Future*' (1987) introduced the now fashionable concept of sustainable development, his first research results showed that indigenous agricultural systems have often used available resources more wisely over thousands of years than most modern agricultural methods do today.

As the recent intensification of agricultural practices based on high capital investment and purchased fertilizers and pesticides has created both economic and environmental problems, the search for alternative forms of agriculture has paved the way for increased scientific attention to indigenous agricultural knowledge and practice. It is fascinating to see how Donald Innis' work on intercropping is contributing to meet the current objectives of such future forms of sustainable agriculture, and in what ways intercropping practices are more advantageous than monocropping in terms of maintenance of biodiversity, risk avoidance, alternative nutrient cycling and weed control management, alternative pest and disease management and improved yield production and soil conservation. His meticulous analysis of the scientific base of different traditional forms of agriculture is leading the search for integrated systems of indigenous and modern agriculture, such as sustainable agriculture and agro-ecology.

In the international debate, the study of small-scale farming systems has now become most relevant to the current strategy of Sustainable Agriculture and Rural Development (SARD), as most food production in the tropical regions of sub-Saharan Africa, Asia and Central America is largely provided by small farms. Here, three-quarters of all farms are on average less than 5ha in area. The practice of growing more than one crop in a field in a year, generally referred to as multicropping, as a widespread feature of these small farms has received increasing attention from agricultural scientists and extension workers, and the number of articles in scientific journals, books and manuals on intercropping has grown substantially during the past few years.

In order to illustrate further the intensity and range of such interest, David Brokensha (pers. comm. 1996) – the Nestor of the study of indigenous knowledge systems in the context of development – has provided us with a timely synopsis, reproduced here, of five selected journal articles, or chapters in books, all of which were published in the 1990s. These are, to some extent at least representative of the large number of recent publications on this subject. In line with Donald Innis' approach, it is noteworthy that nearly all five publications try to adopt the perspective of the small farmer, stressing his (and often *her*) significance in the overall scheme. Charles Francis (1992: 94) recommends a research agenda for the relevance of intercropping to sustainable agriculture which would include:

- alternative nutrient sources and nutrient cycling;
- cropping systems research;
- alternative weed control strategies;
- alternative disease and nematode management strategies; and
- alternative insect management strategies.

Note the emphasis on 'alternative strategies', implying a dissatisfaction with many mainstream practices. As Francis (1992: 94) states: '. . . future strategies for research on intercropping need to take into account the information already available, both from research stations *and from farmers*' (emphasis added). The last three words are important, for it is vital to continue the research '. . . on the links between researchers, extentionists and farmers . . . to help improve the productivity of the intercropping system'.

Anan Pol Thanee (1992) is also concerned, as the title of his article indicates, with 'Farmer as Scientist: Bringing the Farmer's Knowledge to Research'. He provides examples of unusual mixed intercropping from Northeast Thailand: onions and peppers are grown together, as are peanut or mung beans with maize. Thanee (1992: 150) recognizes both the value of farmer knowledge of such practices, and also: '. . . the barrier of acquiring a better understanding of this knowledge . . . by agricultural scientists . . . who need to develop the skills . . . to collect this information and work our way round vague answers and farmer reluctance to answer direct questions . . . to make our own work more successful we need to take full advantage of any opportunity to work with them *and learn from them*' (emphasis added). Researchers in this area need to recognize that their role includes learning from farmers, as well as instructing them.

The next two examples concern specific case studies. Clifford Gold (1993) looks at cassava intercropping in Latin America, with particular concern for plant growth and resistance to pests (whitefly). He concludes that: '. . . crop pests have demonstrated a wide range of responses to diversified cropping systems . . . the tendency is for intercropping systems to reduce levels of specialist herbivores'. Gold (1993: 138) points out that: '. . . not all intercrop combinations are equally effective at reducing cassava herbivore levels and increasing yields'; nevertheless, he believes that his studies show that: '. . . manipulation of cropping system provides an ecologically based front line of defence which can serve to discourage herbivore buildups . . .'.

The second case study, by Ricardo Godoy and Christopher Bennett (1991) examines monocropping and intercropping of coconuts in Indonesia. Here, although intercropping generates more income than monocropping, donors and governments still encourage monocropping. As well as being profitable for smallholders, intercropping: '. . . improves soil fertility, water retention and soil temperature, controls pests and diseases and enhances agricultural productivity . . .' and results in '. . . unambiguously higher financial and economic returns than monocropping'. Godoy and Bennett (1991: 94, 96) base the monocropping preference of development agencies and governments on two factors: first, more attention is paid to *yields* than to smallholder income; and second, it is easier to monitor monocropping than intercropping. 'The bias against intercropping favours the better-off farmers who can afford to set aside tracts of land for monocrop production . . . restrictions on intercropping produce regressive income effects in the countryside'.

Finally, Peter Martin Ahn (1993) writes a short but important article on 'Soil Management in Multiple Cropping Systems'. While reiterating that: '. . . multiple cropping has received increasing attention from agricultural scientists in recent years' the author examines different forms of multiple cropping systems, noting that: '. . . multiple cropping has many forms reflecting different ways that two or more crops may be combined (intercropping systems), or succeed each other on the same piece of land in the same year (sequential cropping systems)'. He observes (1993: 222) that intercropping often improves soil management and produces greater yields because: '. . . the mixing of crops gives a better utilisation of space, time and the resources of sunshine, water and soil nutrients'. However, the greater yields are often '. . . probably of less concern to many tropical subsistence farmers than risk avoidance'. Many anthropologists and agriculturalists have also drawn attention to this crucial (for poor peasant farmers) factor of avoidance of risk.

Recently, these largely economic arguments in favour of such multiple cropping systems have been enhanced by Bray's (1994: 21) study of intensive polyculture, exemplified by rice cultivation. She not only notes that monocropping has become more vulnerable to crop pests and price fluctuations, but that new monocropping technology uses enormous amounts of costly chemicals and fossil fuel: 'In energy terms, it is less efficient than many traditional farming systems'. Moreover, she illustrates that comparisons of agricultural productivity of monoculture and polyculture generally put the latter into an unfavourable light, since such typical economic calculations of productivity are only based on yields of one single crop, ignoring other yields of the polyculture. 'Real' measurement of overall productivity shows much higher output in mixed farming systems than in monocropping systems.

While most of these studies are seeking to highlight the economic dimension of multicropping systems, the recommendations of 'Agenda 21' of the UN Conference on Environment and Development (UNCED), held in 1992 in Rio de Janeiro have given fierce momentum to the interest in the ecological dimension of biodiversity of intercropping systems. During the Earth Summit, the emphasis was placed on 'biodiversity', shifting the attention to the new concepts of 'sustainability' and 'conservation' as alternative approaches to the dominant strategies of 'development' and 'growth'. Since Western-oriented agricultural development is largely based on high-input in a limited number of crops for the world markets, it has created severe problems of reduction in biodiversity as well as in economic and cultural diversity over the past decades. The obvious advantages which intercropping systems are providing to the conservation of biodiversity in terms of genes, species and ecosystems necessitates further study and documentation in order to attain – as Shiva (1996) describes – a biodiversity-based agricultural productivity framework in which diversity is well taken into account

It is indeed gratifying that these leading articles all, in different ways, underline the continuing significance of Donald Innis' pioneering studies, and also point to the need for extending research in the vital area of intercropping, with its potential benefits for agricultural sustainability and production, and for conservation of biodiversity. As such, this work provides *the source book* for the further development of the theory and practice of intercropping systems and sustainable agriculture, and their implications for food shortage reduction in the tropics.

As the extensive bibliography of Donald Innis' studies is basic to the increasing interest in the study of intercropping and traditional agriculture shown by extension workers, development experts and agricultural scientists participating in workshops, symposia and conferences currently being organized by international research centres and development programmes, it has been updated with the most relevant literature published since he finalized the manuscript shortly before his premature death in the late eighties.

The conclusion that intercropping offers both agronomic and ecological advantages as well as economic and financial benefits over monocropping is well documented by this book and sets the stage for a priority research agenda for intercropping systems in the context of indigenous knowledge and sustainable development for the years to come.

Dr L. Jan Slikkerveer
Associate Professor of Anthropology
Chair LEAD Programme
Leiden University
The Netherlands
July, 1996

1 The nature of intercropping

Introduction

HOW DID THIRD WORLD farmers manage before the First World existed? Did people suffer often from mass starvation and the soil from massive erosion? Was the soil of the Third World countries in better condition when First World people first arrived than it is now? How did the farmers manage to maintain soil fertility for thousands of years without the use of non-renewable resources? This book attempts to answer this last question, and to suggest answers to the others, by examining intercropping, a major cultural achievement of Third World peoples but one which is seldom practised by modern farmers.

The efficiency of Third World traditional agriculture has been little studied during the period of contact between European and other cultures. The general assumption, on the part of peoples of European origin, is that European methods are superior and that there has been no real need to study those of others. In the late twentieth century, however, the costs of modern agriculture are rising, as is the damage caused by erosion, so it is not surprising that research on traditional methods is now being undertaken.

This study of intercropping begins with a survey of the main conclusions reached by modern scientific research; Chapters 2–4 are devoted to case studies of traditional intercropping patterns in three different parts of the Third World; while in Chapters 5 and 6 an analysis of the data and findings of the earlier chapters is attempted.

The development of agriculture over the last 10 000 years has involved not only the selection and breeding of wild plants to form productive crops but also the development of methods of planting which maximize yields and maintain soil fertility. Indigenous agriculture all over the world commonly makes use of intercropping, the practice of growing two or more crops at once in the same field. The worldwide distribution of the practice would seem to indicate that it is an ancient cultural invention. In each region, a group of crops derived from local wild plants has been domesticated, providing a balanced diet. In the Americas, the principal grain and legume crops are maize and beans, in Africa millet and cowpeas, in the Middle East wheat and chickpeas, in India sorghum and pigeon peas (plus many other legumes), in China rice and soybeans, and in Europe oats, peas, beans, and barley. These crops are adapted to local environments and also to the other intercrops with which they are grown.

This study is not concerned with the domestication of plants (see C.O. Sauer, *Agricultural Origins and Dispersals*, 1952) but with the ways in which plants are fitted into intercropping combinations. The study demonstrates how indigenous agricultural methods, developed gradually over periods of thousands of years, have produced methods of growing crops which use resources more wisely in the short run and more profitably in the long run than do modern mechanized systems. It is not primarily concerned with how advanced civilizations of the old and new worlds have created deserts and treeless areas, but with how small rural farmers have kept soil fertility at useable levels for centuries, or millennia, and have, as we shall see, produced a balanced diet while using resources in a non-destructive way. With the increasing difficulties encountered by modern agriculture in terms of soil erosion and the depletion of non-renewable resources, it becomes important to examine the effectiveness of traditional farming methods.

Ecological advantages of intercropping

Intercropping will probably be an essential part of future agriculture because it uses environmental and other resources more efficiently than does monocropping (the practice of growing a single crop in a field). Only those types of agriculture which maintain soil fertility can survive in the long run. Monocropping, which often allows erosion to proceed more quickly than soil can be formed, and which allows many other kinds of environmental damage, has clearly a limited future.

Agricultural planting and harvesting methods can be compared to forestry practices to explain why intercropping is more productive in terms of yield for a given energy input than monocropping and less destructive in terms of environmental impact. The use of intercropping can be compared to the selective cutting of trees, the use of monocropping to clear-cutting.

A natural forest in a temperate climate gradually builds up the soil because photosynthesis by several kinds of trees growing at once in a given area produces organic material faster than it is removed by decay, leaching and surface erosion. When a forest has been clear-cut, leaving only plants which are one metre high or less, the amount of photosynthesis that takes place is much less than that which takes place in a forest where only some of the trees have been cut. Most of the sunlight which falls on a clear-cut forest is wasted since there are no longer several storeys of leaves to use it; in addition, much of the rainwater which falls on the deforested land is lost because the small root systems are likewise unable to use most of the water. Much less biomass is produced after a forest has been clear-cut or after a monocrop has been harvested; consequently, much less organic matter is added to the soil.

Not only does less growth and less soil-building take place in a region from which the forest or crop has been removed, but the environmental factors which are no longer being used for growth become destructive. The sunlight which is no longer captured by leaves heats the ground, causing more water to be lost by evaporation and accelerating the decay of organic matter. Since few roots are left to utilize the elements made available by the decay of organic matter, much of the nutrient material is washed away over the surface or is leached downward through the soil out of reach of the small plant roots. There is also more erosion, since falling raindrops are not slowed down by the numerous leaf surfaces of a forest or a field full of different crops. Less water is held by spongy decaying leaf layers on the ground or by organic materials in the soil since these both diminish as decay continues, and fresh organic matter is not added at an adequate rate. Since less water is held by leaves and stems on the ground and in the spaces of the crumb structure of a good organic soil, there is more water to run away over the surface and cause erosion, and more water leaches down through the soil, carrying soluble nutrients away from the root zones.

Because considerable vegetation remains after a partial harvest, there is no period of rapid soil loss through decay, erosion and leaching. The parallel between intercropping in traditional agriculture and selective cutting of forests also holds true for economics. There is a real difference between farming or logging for maximum profits and farming or logging for maximum yield. In modern economics, maximum profits from a business operation are usually defined as short-term profits. There seems to be no satisfactory way in which the free market can allow for the fact that maximum profits in the present often involve destroying resources and making future maximum profits much smaller. Clear-cutting of forests is more profitable because the equipment and crews employed to harvest the trees can cut more lumber in a given time and spend less time travelling from one tree to another. But in the long run, as we have seen, clear-cutting is destructive and probably not profitable because it allows the erosion and leaching away of resources which would have remained available for trees in a natural forest or in a selective cutting operation.

Since many people understand the dangers of clear-cutting forests, there is some social pressure for lumber companies to behave in a more socially responsible manner from a long-term point of view. In some countries forests are cared for, replanted with several different kinds of useful species, and harvested selectively. Relatively few people, however, are aware that the same arguments apply to agriculture. Agricultural research on the benefits of traditional and modern intercropping has, unfortunately, been very much neglected. It is widely accepted in Europe and America that the best method of producing food is to monocrop. In Third World countries, the lack of scientific understanding of intercropping has meant that many skilled traditional farmers who looked after the land with care are being replaced by fewer, more mechanized, monocropping farmers who allow environmental destruction for short-term profit.

Fieldwork on traditional agriculture, as will be shown in later chapters of this book, shows that intercropping is very common and extremely sophisticated within the traditional cultures. Experimental work at agricultural research stations is beginning to measure the efficiency of traditional farming practices in utilizing resources. Using the same resources, intelligent intercropping can produce a greater yield of organic matter, or food, than monocropping can; by intercropping with legumes, the traditional farmer adds nitrogen to the soil. Numerous experiments have shown that with given amounts of water and light, greater growth is possible with intercropping than with monocropping, where all other factors are equal.

Effect of intercropping on yields

A simple experiment which can be done in either temperate or tropical lands shows why traditional farmers have been reluctant to use modern methods. If one uses three equal-sized plots of ground with monocropped beans in the first plot, monocropped maize in a second, and intercropped maize and beans in the third, it is easy to measure the benefits of intercropping and to see how the fast-growing young bean plants provide soil cover and use sunlight which the maize is not yet ready to use. The two plants have quite different growth habits. The more or less vertical leaves of maize trap about half the available sunlight, and the lower more horizontal leaves of beans can capture most of the remaining solar energy. In one such experiment in western New York State, 99 bean plants in 4.5 square metres ($4.5m^2$) produced 708 pods for use as a green vegetable. The green beans were picked several times a week as individual pods became large enough to eat. So, many more beans are produced for the same resources than would have been produced in a commercial operation where green beans are picked only once by machine, leaving the plants so badly damaged that no more can be harvested from the same stock. The contrast between picking a few beans every day or so for weeks, and picking all the beans that happen to be ripe on one day, is a clear example of the difference between maximum yield and maximum profits.

In the New York monocropped maize plot, 33 maize plants produced 38 maize cobs. The intercropped plot had 33 maize plants but only 66 bean plants because one bean plant was omitted from the pattern for each maize plant included. The yield from the intercropped plot (average of three years) was 490 bean pods and 25 maize cobs. An important point to be noted, as can be illustrated here, is that the results of such experiments can be used in arguments both for and against intercropping. The proponents of monocropping can argue that the yield of each individual crop is less when intercropped; thus, the two crops must be competing for nutrients, water and light. The proponents of intercropping, however, can point out that the total yield of the two crops is greater when two crops are grown. If there are two intercropped plots ($9m^2$) making the intercropped area equal to the monocropped area, the total yield would be 980 bean pods plus 50 maize cobs instead of only 708 bean pods and 38 maize cobs which the same resources and the same area produced under monocropping.

The International Rice Research Institute (IRRI) of the Philippines has developed an effective method for comparing monocropped and intercropped yields. It is the land equivalent ratio (LER). In the example above, the intercropped beans produced 0.69 of the yield of monocropped beans, while the intercropped maize produced 0.66 of the yield of monocropped maize. When these are added together, the total LER is 1.35. This means that the same resources of sun, water, nutrients and space produced 35 per cent more when the two crops were intercropped than when each was monocropped. Many experiments have been done recently which measure the benefits of intercropping. Appendix 1 lists several hundred LERs derived from experimental data from many countries; there is no longer any reason to think of intercropping as an untested method. In Appendix 1, experiments are arranged by crop species, giving crop varieties, location, row arrangement, plant-density yields, proportion of intercrops, planting patterns, and both monetary and physical LERs where the information is available or where it can be calculated. Most of the experiments cited compare the monocropped (MC) and intercropped (IC) yields of only two crops. When the yields of three and four crop mixtures are measured, in imitation of the planting methods of many traditional farmers, the yields from the same resources can be two or even three times the yield from monocropping. In fact, the inefficient use of resources which has accompanied the change-over from indigenous intercropping to modern monocropping may be a major cause of world hunger.

Factors involved in intercropping

Through the use of representative experiments the following sections show how researchers have measured some of the numerous factors which are involved in intercropping.

Weight
An experiment in Wales with 36 plants per pot consisted of cutting off the above-ground parts of the plants and weighing them four, eight and sixteen weeks after planting (Fig. 1).

The crops used were barley, mustard and poppy. When all 36 plants were the same species, the biomass weight was less than if two species were grown together (18 of each = 2 IC). Biomass weight was the highest when the three

	Weight of 36 plants		
	31C	21C	MC
4 weeks	16	15	10
8 weeks	42	38	28
16 weeks	53	48	38

(grams)

Barley, mustard, poppy

Fig. 1: Effect of intercropping (IC) and monocropping (MC) on rates of biomass growth (Wales)
Source: Haizel, 1972

species were intercropped together (12 of each = 3 IC). Different root systems and different leaf systems are able to harness more light and make use of more water and nutrients than is the case when the roots and leaves of only one species are present. When only one species is grown, all the roots tend to compete with each other since they are all similar in their orientation and below-surface depth. Similarly, the leaves of plants of the same species are directly opposite and growing at the same rate as each other, whereas the leaves of a plant of another species do not compete so directly for sunlight in space or time. One wise researcher in India summed this up by writing that the worst thing to plant next to a maize plant is another maize plant.

Spacing
In Illinois, oats and soybeans were planted in alternate rows with different spacings between the rows (see Figs 2–5). In an arrangement of rows of oats 21cm apart with soybean rows halfway between, the oats gave 0.95 of the yield of monocropped oats, and the soybeans gave 0.47 of the yield of the monocropped soybeans at the same spacing. Thus, the LER was 1.42. With spacing between the oat rows of 41 and 61cm., the LER was 1.62, and a spacing of 81cm gave a yield 82 per cent higher than did monocropping (LER 1.82) (Chan and Brown, 1980).

Figs 2 to 5 show not only that the benefits from intercropping increase as the plants have more room (which also means more light, water, and sunlight per plant), but also some of the more complicated aspects of intercropping. The linear graphs below each spacing diagram show that the greatest combined yield of oats (o) and soybeans (s) (6000kg/ha) was in the field with 41cm spacing between oat rows rather than in with 81cm spacing, which was the field with the highest LER. It is evident, therefore, that space and resources were being wasted when the distance between the oat rows was more than 41cm. The greatest yield of oats by weight was in the field with the closest spacing, while the greatest yield from soybeans was in the field with the widest spacing. Maximum benefits from intercropping, then, come when the declining yield from more widely

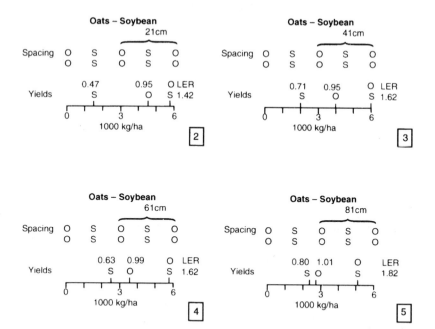

Figs 2–5: Effects of changing spaces between the rows on intercropping yields (USA)
Notes: o = oats, s = soybeans. In Fig. 2, 0.47 means that intercropped soybean yield was 0.47 of monocropped soybean yield; 0.47 plus 0.95 of monocropped oat yield gives total yield of 1.42 where monocropped yield would be 1.00 for either crop.
Source: Chan and Brown, 1980.

spaced oats is more than offset by the increased yield from more widely spaced soybeans.

In intercropping studies, it is always important to note actual yields of the crops by weight and not just the benefit due to intercropping (the LER). Oat rows 81cm apart, for example, are not at the optimum spacing for oats; thus, comparison of monocropped oats at 81cm with intercropped oats at 81cm does give the highest ratio (1.82), but it does not give a farmer the highest oat yield per hectare since oats can give a much better yield if the rows are closer together, for instance, 21cms.

LER variations
Fig. 6 suggests how much the LER can vary in different experiments. The graph is based on 42 experiments with maize and soybeans conducted in different places which, of course, means different climates and soils. The experiments also used different varieties of maize and soybeans.

Many intercropping trials are now being carried out with crop varieties (i.e. cultivars or cultigens) which were bred for highest yields as monocrops. This may explain why two of the experiments gave 10 per cent less yield for inter-cropping than they did for monocropping. Hybrid maize has now become so productive that any reduction in the number of maize plants, or reduction of

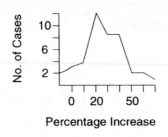

Fig. 6: Variation in intercropping yields (various countries)

maize yield through competition with another crop, may be difficult to offset by extra yield from a second or third crop. The examples shown in Fig. 6 indicate that intercropping maize plus soybeans usually gives higher total yields than when either one is monocropped, where all other factors are equal. There were twelve cases, for example, in which the LER was 1.20 (a 20 per cent increase) (Abraham, 1974; Ahmed, 1979; Alexander and Geuter, 1962; Beets, 1975; Beste, 1976; Cordero and McCollum, 1979; Crookston, 1976; Dalal, 1977; Finlay, 1974; King *et al.*, 1978; Pendleton, *et al.*, 1963; Raynolds and Elias, 1980; and Reddy and Reddi, 1981).

Nitrogen (Figs 7 and 8)
There have been some problems in assessing the role of nitrogen (N) in intercropping. Legumes can fix atmospheric nitrogen and make it available to other plants, but they tend not to do this if mineral nitrogen fertilizer has been added.

Figs 7 and 8, from Cordero and McCollum's North Carolina experiment, provide an example of this. Some researchers automatically add nitrogen fertilizer in intercropping experiments because all modern agriculture uses mineral fertilizers, but this addition of N reduces the benefits to be derived from the practice of intercropping. This has led to some reports that legumes are not beneficial in maize fields. When no N was added (Fig 7), the LER for maize and soybeans was 1.68, but when a significant amount of N was applied, the LER was only 1.33. However, the total weight of grain is much higher when nitrogen is applied. Even so, intercropping, when fertilizer is applied, is still very worthwhile because the total yield is 33 per cent higher. In cases where the cost of fertilizer is very high or where it is not available, intercropping can be very important. It is unlikely that the soybeans (Fig. 7) provide much N directly to the maize plants, but at least the soybeans provide themselves with N and are not competing with the maize for this crucial element. In the plots where no mineral N was added, the intercropped maize yield was 1.09 times the monocropped yield. When N is applied (Fig. 8), the soybeans do not fix atmospheric nitrogen and therefore start to compete with maize for nitrogen. When N fertilizer was added, the yield of maize was higher, the competition from soybeans was greater, and the intercropped maize yield was only 0.84 of the fertilized monocropped maize yield. The reduction in the benefits of intercropping when mineral nitrogen fertilizer is added has been noted by several other research workers

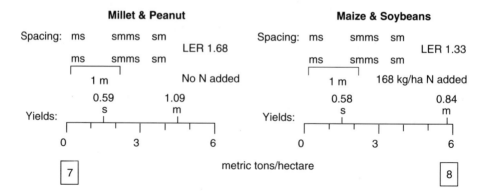

Figs 7, 8: Effect of added nitrogen on intercropping yields (USA)
Note: m = maize, s = soybean.
Source: Cordero and McCollum, 1979.

(Bhalerao, 1976; Enyi, 1973; Fisher, 1979a; Greenland, 1975; Koregave, 1964; Lipman, 1912; Malithano and Van Leeuwen, n.d.). One scholar commented that when N is added non-legumes make just as good intercrops with grains as do legumes (Kurtz, 1952).

Much recent research on the benefit of intercropping with legumes has been inspired by the high cost of nitrogen fertilizers. In Sri Lanka, the addition of 25kg/ha of N made intercropping 51 per cent more profitable than monocropping, but when no N fertilizer was added, the net yield (fLER) was 81 per cent more for IC than for MC (Gunasena, 1980).

In a South Indian experiment, the addition of N helped the pearl millet, but this in turn hindered the growth of the intercropped peanuts because these were more shaded by tall millet than they would have been by the shorter non-nitrogen fertilized millet. When no N was added, the yield of millet was reduced from 8 000 to 5 000 kg/ha but the yield of peanuts increased since these were shaded less by the millet. It happened that the amount of light intercepted was the same in both cases since the decline in millet leaf area was just balanced by the increase in peanut leaf area (Willey and Reddy, 1981). As early as 1912, Lipman found in New Jersey experiments that adding nitrate to soil made maize and oats grow so tall that they outgrew and smothered intercropped legumes (Lipman, 1912). In Illinois, a tall variety of sorghum resulted in less nitrogen being fixed by intercropped soybeans than was the case when short varieties of sorghum were used (Wahua, 1978). The development of cultivated varieties more suitable for intercropping, or the rediscovery of old varieties, would seem to be a fruitful line of future research for plant breeders.

Water
Different species of plants do compete for water when it is scarce, but two crops can recover more of the available water than one. A variety of kinds of root systems reduces water loss which in turn reduces loss of nutrients through leaching.

In Figs 9 and 10, when there was no shortage of water (Fig. 9), the LER was 1.18, and the yield of the IC millet was about 1200kg/ha or 0.40 of the monocropped millet. When there was not enough water for optimum growth (Fig. 10), the total yield was much less, but each crop was nearer the monocropped yield in times of drought, giving an LER of 1.30. The extra root systems capture more water, thus more water is transpired which tends to keep the local environment cooler. The extra shade also cools the soil, helping to create a micro-climate with lower temperatures, less evaporation, and higher relative humidity (Reddy and Willey, 1980). If one crop is much more valuable than the other, the reduction in yield of the main crop by the intercrop will make intercropping appear undesirable. This has been noted with a cover crop in Nyasaland (now Malawi) tea plantations (Laycock and Wood, 1963) and during the short rainy season in Tanzania where beans reduced the maize yield (Fisher, 1977). During the long rainy season in Tanzania, however, intercropped maize and beans achieved an LER of 1.30.

In India, more water was used when secondary crops were grown with maize or sugar (De and Singh, 1979), which would be a profitable way to plant the field if more water was available than could be used by the sugar. In another experiment in India, however, the growth of a second crop did not increase the use of water. Monocropped jowar (sorghum) and intercropped jowar plus pigeon peas required the same amounts of water, apparently because the water which would have evaporated from the ground was channelled through the pigeon peas instead (Nataranjan and Willey, 1980). Intercropping can use water more efficiently if there is more water than one crop can use at one time but not enough water or a long enough growing season for two consecutive crops (Bhalerao, 1976). A mulch of dead clover can increase maize yield by preventing water loss from the soil (Standifer and Bin Ismail, 1975).

Intercropping has been shown to reduce evaporation from soybeans in Minnesota. Windbreaks of maize increased crop yield by 20–25 per cent

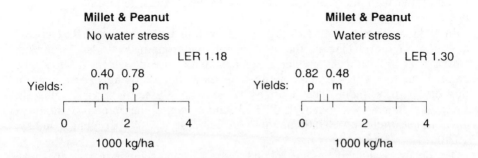

Figs 9, 10: Intercropping and efficiency of water use (India)
Note: m = millet, p = peanut.
Source: Reddy and Reddi, 1981.

(US$12–$15/acre) (Radke, 1968). Intercropping produces more organic matter than monocropping, and additional organic matter in soil increases the soil's capacity to hold water. Unmulched monocropped maize in Nigeria produced a soil which allowed water to infiltrate only one-quarter as fast as in a field which had mulch or an area in which brush was regrowing during a fallow period (Juo, 1977). In Orissa, India, it has been shown that monocropping with a non-legume steadily reduces the organic content in the soil which in turn affects adversely the quality of the soil structure and its ability to hold water. This leads to a decrease in the amount of water stored in the soil for plants to use during dry periods and increases run-off of erosion and leach water, reducing the amount of soil and soil nutrients available. After four crops of monocropped rice, the proportion of water-stable aggregates (particles) in the soil larger than 0.5mm in diameter was reduced from 47 per cent to 14 per cent. Potatoes and peanuts were found to build up water-retaining crumb structure in the soil, while maize and jute reduced it. Some crops favour bacteria which produce mucus which binds soil particles together, allowing water to be stored in the spaces between the particles. This soil is porous, allowing water to infiltrate, but since the mucus at the same time holds on to the water, it does not escape from the soil (Sadanandan and Mahapatra, 1974).

It is widely agreed that small farmers intercrop as a form of insurance against total loss of their crop. It is not widely accepted, however, that small traditional farmers intercrop because it is a better method of farming which gives higher total yields and is better adapted to a potential range of many actual climate conditions than monocropping. In Rajasthan, India, it was found that cluster bean and moth bean gave better yields when the rainy season was long, but that greengram and cowpeas gave less yield in these conditions because of the difficulty in harvesting them. These four legumes were usually used as intercrops with bajri (pearl millet) in the author's Maharashtra study area (see Chapter 4). Thus, small farmers who usually have four or five intercrops in a field are better prepared for whatever the weather may bring than are big farmers who monocrop and who may have chosen the wrong crop for the prevailing weather. Research intended to aid big farmers tends towards manipulation of the environment to make it similar to the controlled conditions of an experimental plot. The traditional farmer with intercropping uses a system better designed for the uncertainties of the real world.

In 1973 in Rajasthan, the average yield for four monocrops was Rs (rupees) 0.98/mm (of rain)/ha, while the average for intercropping was Rs4.33/mm/ha, which is a 342 per cent greater efficiency in the use of water. It should be pointed out that grasses were involved in these intercropping experiments in this dry part of India. The legumes helped the grasses and vice versa. In 1975, the monocrops averaged Rs3.23/mm/ha while the intercrop mixtures gave a net yield of Rs6.06/mm/ha, an increase of 88 per cent (Daulay, 1978).

To obtain water for monocropped hybrids, many modern farmers have become dependent on pumping up water from groundwater supplies. Instead of relying on intercropping and organic soil to hold water for crops, some modern farmers in dry areas, therefore, lower the water-table until it becomes relatively inaccessible to many people (depending on their ability to pay for expensive energy).

Light
An experiment in North Carolina shows how intercropping uses solar energy more efficiently.

The first graph (Fig. 11) shows the Leaf Area Index of maize and soybeans grown alone. The index is proportional to the area of leaf surface which a plant has available to carry on photosynthesis at any one time. At this latitude the maize uses only part of the growing season, and a month or two of late-season sunlight is wasted. The soybeans, on the other hand, which are small at first and slow-growing, waste solar energy in the early part of the growing season. The uppermost line in Fig. 12 shows the leaf area and hence the amount of solar

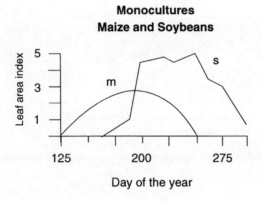

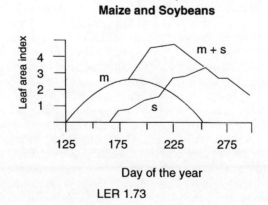

Figs 11, 12: Efficiency of solar energy use (USA)
Source: Cordero and McCollum, 1979.

energy used by both crops. The soybeans have less leaf area when grown with maize, but after the maize has been harvested and the beans are in the pod-forming stage, there is no competition from maize, and the total yield is excellent. The LER is 1.73, a 73 per cent increase in yield by weight for the same resources (Cordero and McCollum, 1979).

Plants of different heights will make more use of light when intercropped than when monocropped. In the tropics, multi-storey plants harvested in sequence can utilize the sun's energy on a year-round basis (Sebastiani, 1981; Osiru and Kibira, 1979). The lower leaves must, of course, be adapted to lower light conditions. In Kerala, India, young areca palms are aided by the shade of bananas. The bananas produce a substantial extra yield before the areca is mature. Elephant foot yam and pineapple under areca palms are profitable intercrops for the small farmers who are the principal growers of areca (Khader and Anthony, 1968). If the leaves of the taller plant are vertical, as in the case of cereals, and leaves of the lower plant are horizontal, making more use of the dim light as in the case of legumes, it is possible to utilize twice as much solar energy with intercropping as can be done with monocropping (Ojomo, 1976; Brougham, 1958; Baker and Yusuf, n.d.). In intercropped fields where the number of plants of each species is equal to that in monoculture, the upper portion of the taller crop which is above the leaves of the lower crop does not suffer from competition for light, and the shorter crop can benefit later from full sunlight if the tall crop is the first to be harvested (Baker, 1979a). If replacement intercropping is practised, in which some individual plants of one species are replaced by plants of the other crop, then the upper part of the tall crop (the part above the leaf canopy of the shorter crop) may be at only half the monocrop density. The tall plants thus have more space and light than they would if monocropped and grow bigger than those which are more closely spaced. If the upper parts of the tall plants are at wide spacing in the first part of the growing season (because they are above the short crop), then they will produce much more per plant. In a Nigerian experiment, millet at half the normal population yielded almost as much as the normal population in the same space with the same resources. After the tall plants are harvested, if short plants such as a long-term legume are still in the field, they will benefit from having more resources per plant and will yield more than plants in a monocrop at normal spacing (Baker, 1979b).

In Tanzania, a maize and bean mixture captured 13 per cent more light than monocropped maize and 6 per cent more than monocropped beans (Fisher, 1976). In Illinois, double rows of maize gave higher yields than the same number of plants per hectare uniformly spaced, but the uniform rows were better for machine cultivation. The double rows allowed soybeans to be intercropped more easily. The maize row next to the soybeans naturally received more light than a maize row flanked by maize on both sides and yielded 25 per cent to 39 per cent more grain. On the other hand, a row of soybeans next to maize received less light than a soybean row with soybeans on both sides, lowering soybean yield from 28 per cent to 34 per cent when it was shaded by maize (Pendleton, 1963). The researcher suggests that another variety of soybean or a different legume might have done better when intercropped with maize. Similar experiments in North India produced comparable results (Mohta, 1980).

The leaf crown on coconut palms in Kerala, India, occupies a space 7.5 × 7.5m in area, but the roots occupy only one-quarter of this area, which means that there is ample room for small farmers to grow secondary crops for food or for additional income. As the leaf canopy allows 50 per cent of the light to get through, shade-tolerant plants can grow beneath it. One hectare of coconuts yielded not only 17 500 nuts in a year but also 100kg of black pepper, 750kg of cacao, and 5000kg of pineapple (Verghese, 1976). Cacao alone, as an intercrop with coconut, can also give excellent extra returns (Nair, 1977).

Sugar cane in the first three months of its growth uses very little of the light energy available. In monocropped cane 70 days after planting, 82 per cent of the solar energy can still reach the ground, whereas if sugar cane is intercropped with mung bean only 60 per cent of the light reaches the ground; with cowpea 57 per cent and with dwarf castor bean 48 per cent of the light reaches the ground (De and Singh, 1979). Sunflower is another tall plant which can easily be intercropped. In Rajasthan, it was found that low legumes such as greengram, cowpea, moth bean and peanut did not interfere with the normal growth of sunflower and gave an average extra yield by weight of 74 per cent (LER 1.74). Cluster bean, an upright legume, gave too much competition with sunflower to be profitable (Singh, K.C. and Singh, R.D., 1977).

In addition to the technique of employing plants of different sizes, it is possible to make greater use of solar energy by planting crops that grow at different rates and that make maximum use of sunlight at different times. In Tanzania, string beans grew faster than maize and reached their maximum leaf area in 36 days. Seventy days after planting, the maize reached 50 per cent of its maximum leaf area, by which time the beans were almost finished (Fisher, 1979c). In other combinations, the cereal may reach maturity and die first, leaving more light, water and nutrients for the legumes. Since cereals are determinate crops, having a certain fixed lifespan, it is possible to choose a cereal variety which will automatically start to produce less shade at a certain time as the seed head ripens. This type of cereal can be an excellent intercrop for legumes which are indeterminate and can keep on growing and producing pods as long as the rains continue (see Genest and Steppler's (1973) Canadian study).

Maize and upland rice are a very common mixture in Asia. An experiment in the Philippines with 90-day maize and 120-day rice resulted in a low yield because the period of growth of the maize overlapped too much that of the rice. In Indonesia, however, a 75-day variety of maize is used which is ideal for intercropping with rice. For the first eight weeks, there is very little competition between the two crops for light, water or nutrients. This is followed by a period of severe competition. The maize is then harvested, so that the rice has the field resources to itself during the period of grain formation. The total weight of the plants (biomass) under different spacings and row arrangements give LERs of 1.5, 1.2, 1.3, 1.6, 1.4 and 1.6. Grain yields arranged in the same order give LERs of 2.6, 1.7, 1.6, 2.2, 1.8 and 1.8. One conclusion which can be drawn is that smaller plants do not necessarily yield much less grain. It is quite common in intercropping experiments where suitable (usually traditional) varieties are used to find that stunted growth of one of the plants need not reduce grain yield. In the Philippines, the lowest densities of plants per hectare had the highest LERs, but these were not the experiments with the highest grain yields. In the wet season, the maize grew well, and the rice could not successfully compete with more than 20,000 maize plants per hectare. In the dry season, 40,000 maize plants gave the optimum total yield for rice and maize grown together (Sooksathan and Harwood, 1976).

An experiment at the International Crop Research Institute for the Semi-arid Tropics (ICRISAT) in Andhra Pradesh, India, suggests that two crops are not enough in an intercrop combination. As we shall see in later chapters, traditional farmers usually grow more than two crops in a field. At ICRISAT, after 82 days the amount of light intercepted by two crops finally rose to 65 per cent, but then *jowar* was harvested, and the pigeon-pea plants which remained intercepted only 19 per cent of the solar energy. Because the crops were in alternate rows, increasing the density of pigeon-pea plants within the row did little to make more use of sunlight (Natarajan and Willey, 1980). In the Cameroons, a native expert was asked to arrange the crop spacing in his traditional way. When this 'random' non-row pattern of intercropping was compared with row intercropping, it was found that where the method of spacing was the only variable changed, the traditional expert achieved 36 per cent more maize yield, 25 per cent more *macaba* (a root crop), and 31 per cent more *taro* (IRAT, 1969). The traditional farmer used the same techniques that are used by good flower gardeners in Europe and America who seek out suitable garden spots for each different species to produce a continuous series of blooms over a long period of time. When food crops are arranged in this way, there is always a high percentage use of available sunlight, water and nutrients. This yields more food and at the same time protects the soil because more total growth takes place than would occur with a monocrop, where all other conditions are equal.

Time

Arranging long- and short-term plants together in a field can increase efficiency in the use of light, water and nutrients. It is quite common in Nigeria to plant millet, sorghum and cowpeas together. An intercropping experiment using local methods and local plants allowed much more growth to take place in the field each week than was the case where only one crop grew at a time.

As Fig. 13 shows, 80-day millet grows to a height of 1.5m and is then harvested to leave more resources for the slow-growing sorghum to use (Baker, 1979a). This not only enables more growth to take place but provides an early crop for families to eat, reducing the time when the family has to be on short rations at the end of the long dry season.

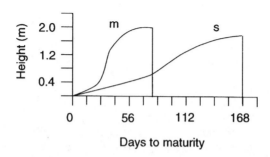

Fig. 13: Intercrops maturing at different times (Nigeria)
Note: m = millet, s = sorghum.
Source: Baker, 1979a.

Figs 14 and 15 show how cowpeas are worked into this traditional Nigerian system of intercropping. After the pearl millet is harvested, cowpeas are planted in the spaces where the millet was.

This experiment used rows of plants because it was designed to show how small machines could be used. However, as we have seen, the native experts obtain higher yields by not planting in rows. Closer association of plants in non-row spacing tends to increase the benefits of intercropping since plants with different leaf architecture can fit one over another more efficiently; also, root systems can interpenetrate and to some extent lie one underneath another to use a variety of nutrients at different times and at varying depths in the soil. Whereas the traditional Nigerian farmers practise relay cropping by sowing cowpeas before the millet harvest, Andrews first harvested the millet and then planted the cowpeas. The long-term sorghum was in the field during the growth periods of both of these other crops. Relay cropping of the cowpeas gives them a head start and enables the small cowpea plants to utilize nutrients and water and protect the soil from erosion immediately after the millet harvest which occurs during the rainy season. Modern agricultural methods tend to ignore the long-term effects of soil erosion; planting methods which have succeeded for a thousand years or more would not have survived if this factor had been ignored. The benefits of growing three crops at once and also of relay cropping can be very great, as the LER of 1.88 testifies. The monetary value of the three crops is almost twice the value of a single crop and is, in fact, so high that even after the cost of the extra labour that intercropping requires is subtracted, intercropping is still the more profitable method (Andrews, 1972).

Now that research workers are becoming interested in the possibilities of intercropping, some quite sophisticated work is being done. Fig. 16 shows how total yield is affected by the length of time between maturity of the two

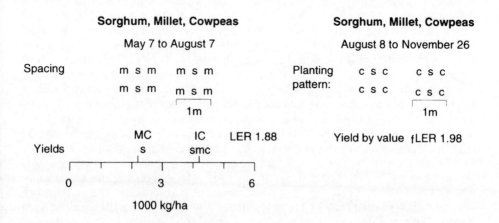

Figs 14, 15: Relay intercropping millet replaced by cowpea (Nigeria)
Note: m = millet, s = sorghum, c = cowpea.
Source: Andrews, 1972.

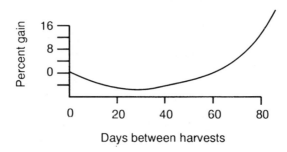

Fig. 16: Competition between sorghum and millet (Nigeria)
Source: Baker, 1979a.

crops. Baker used different varieties of millet and sorghum to test this effect; the graph in Fig. 16 shows the results. When dates of maturity of these two crops were within 40 days of each other, there was no increase in yield due to intercropping. Two crops which are both making their maximum demand on environmental factors and fertilizer resources at the same time will be competing for sunshine, rain and nutrients, and both will suffer in the competition. But if one plant is small (such as young sorghum) while the other is large (mature millet), there is less competition. The millet can use the resources first to produce a good yield. If the date of maturation of the sorghum is more than 40 days after the harvest of the millet, there will be a net benefit from intercropping (Baker, 1979a).

Relay planting is a widely used method of encouraging intercrops to make their maximum demands on resources at different times. Not only are rice and maize intercropped in Indonesia, but cassava is also planted in the same fields. But if the cassava is planted at the same time as the two grains, its leaves and root systems will compete with them too much; it is, therefore, not planted until 30 or 40 days after the planting of rice and maize (relay planting). In this way, the harvest dates are widely spread with the maize harvested after 80 to 85 days, the rice at 140 days, and the cassava after a year or so (Effendi, n.d.).

Where labour supplies are plentiful, as is the case in many poor countries and on farms run by families, sophisticated kinds of hand interplanting can be carried out. In Nepal, finger millet is relay-planted between standing stalks of maize, enabling the millet to ripen while the weather is still favourable. This prevents the soil erosion which would occur if one crop were harvested, the ground dug up, and the second crop then sown. Similarly, in Jamaica, the author has observed that sweet potatoes are relay-planted with yams well before the yam harvest. This enables the sweet potatoes to get a good start and to cover the soil with vines soon after the yam harvest. In Madras, India, it has been found very profitable to plant finger millet in January and then to relay-plant peanuts in March (Pillai, 1957).

In the northern United States, the growing season is not long enough for two crops to be grown one after another in the same year, but wheat and oats do not occupy the whole growing season. Some experiments have been done in which soybeans are relay-planted three or four weeks before the grain is harvested (Jeffers, 1979 in Ohio; Chan and Brown, 1980 in Illinois; Graves, 1978 in Tennessee). This makes it possible for the soybeans to ripen and be harvested before the frost comes. Intercropped soybeans need not hinder the grain harvest if the cutting head of the combine is set high enough to avoid cutting off the tops of the young soybean plants. In order to sow the beans without damaging the wheat or oats, the method often used is to hire an airplane. This is expensive, and if rains are so poor that few soybeans germinate, the practice is not worth while. Although American farmers are reluctant to return to broadcasting seed by hand or by a hand-held machine, in one experiment where hand methods were used to sow soybeans in standing grain, the process was referred to by the morale-boosting phrase, 'airplane simulation'.

Nutrients
Many traditional small farmers all over the world have lost their land because they lacked the vocabulary to explain in scientific terms what they were doing. Agboola and Fayemi (1971) in Nigeria, at least, have shown how intercropping is advantageous in conserving soil nutrients.

Fig. 17 shows the planting pattern for maize and greengram, the yield when no fertilizer was added, and the high LER of 2.08. In this case, maize is the tall plant which benefits (19 per cent extra yield) from having a short legume in the next row instead of another row of maize. Fig. 18 compares percentage nutrient loss from the soil when maize is grown alone, with soil nutrient loss when maize is intercropped with cowpea and with greengram. In other experiments, it is usual to measure the nutrient content of the plants, which can give the impression that if the weight of two crops is more than the weight of one, there must be fewer nutrients left in the soil. The experiments of Agboola and Fayemi show, however, that this is not the case.

In every case except one in Fig. 18, there was more soil nutrient loss when one crop was grown than when two were grown. Legumes supply their own nitrogen when fertilizer is not used, which means that there is usually no more nitrogen taken from the soil by two crops than by one if one crop is a legume. The fact that two crops take less phosphate and potash from the soil than one crop is more unexpected. The explanation is that with only one root system, there is more nutrient loss through leaching and erosion than with two root systems. When the second root system lies partly underneath the first, nutrients can be captured which would be lost through leaching if only the first root system existed. The extra nutrients recovered are brought back up to the surface by the longer-rooted crop. When the leaves and roots of the legume die, the recovered nutrients are returned to the surface layers of the soil. This is one of the most important ways in which intercropping can be more efficient than monocropping. It is sometimes less profitable than monocropping because of the extra labour involved in working with two crops in the same field at once, but even when it is less profitable in the short run, it is more profitable in the long run because it maintains soil and soil nutrients for future generations, or even for the same generation at a later date.

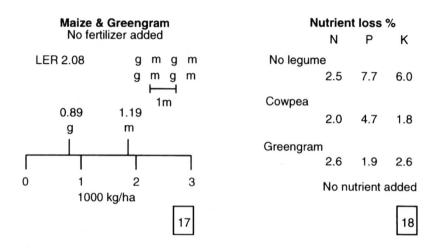

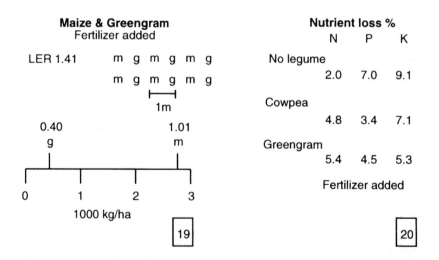

Figs 17–20: Efficiency of nutrient use. Maize monocropped and intercropped with and without fertilizer (Nigeria)
Source: Agboola and Fayemi, 1971.

Figs 19 and 20 show that intercropping can also be very useful when fertilizers are added. The maize gives the same yield with and without intercropping (IC yield is 1.01 of MC yield) since the greengram provides very little competition. The yield of fertilized maize is higher by around 600kg than the yield of unfertilized maize (compare with Fig. 17). The LER is lower (compare Figs 19 and 17) because the fertilizer allowed the maize to grow taller and to deprive the greengram of light, water and nutrients which the greengram would have received if the intercropped maize were shorter.

The table of nutrient loss in Fig. 20 shows once again that legumes do not fix atmospheric nitrogen when nitrogen fertilizer is added to the soil. More than twice as much N was taken from the soil by two crops than by maize alone, showing that the legumes are removing it from the soil instead of from the air as they did in the 'no fertilizer' experiment. the phosphorus (P) and potash (K) loss shows the same pattern as in the previous experiment. If one remembers that soluble nutrients are always being washed down through the soil whenever rain falls, the role of intercropping becomes clear. The cowpea and greengram roots are able to retrieve potash and phosphate which would otherwise have been lost, and return them to the upper layers of the soil. This explains why two crops grown together leave the soil in better condition than one crop grown by itself (Agboola and Fayemi, 1971).

Many people feel that 'modern' agriculture is superior because it has such high yields. Yet these high yields are possible only when extra nutrients are supplied. One reason for this may be that modern monocropping uses nutrients inefficiently. One must conclude that intercropping makes the best use of the available nutrients whether or not these include added minerals. As non-renewable nutrients from mines become scarcer and more expensive, it will become more necessary to use nutrients more efficiently.

Traditional farmers in India have long practised continuous cropping without noticeable detrimental effect to the soil, probably because crop mixtures are used (Wilson and Wyss, 1937). Intercropping can be regarded as instant crop rotation. 'The system of mixed crops, so common in India, is undoubtedly a successful and profitable method which has probably done more to uphold the fertility of Indian soils than any other practice' (Mollison, 1901, quoted by Bains, 1968).

In Taiwan, experiments with rice, sweet potatoes and melons, in which soil fertility was measured before and after the crops were grown, showed that if the recommended amounts of fertilizer were applied to intercropped fields, the levels of available P and K in the soil started to increase but that this was not true in monocropped fields. This is because two crops are able to prevent some of the losses that would result from leaching (Su, 1975). In the Punjab, India, it was shown that intercropping recovered much more N, P and K from the soil than did monocropping (Sharma, 1979).

When the root systems of different crop species are in the ground at the same time they recover water and nutrients more completely as they draw on the resources from different depths. If these nutrients are derived mostly from organic matter which decays gradually and makes nutrients available over time, it is profitable to have several different root systems in place to capture them as they become available and thus capture more of them. Soybeans, for example, have many roots below the root system of maize where they capture nutrients which have slipped through the maize-root hairs (Beets, 1975; Crookston, 1976; Greenland, 1975; Osiru and Kibira, 1979; Sebastiani, 1981). It would seem to follow from this line of argument that three crops grown together would produce a higher yield than two crops. But if the cost of labour is high, monocropping will be the most profitable, but not the most productive, method (Gupta and Mathur, 1964). However, it is only a short-term profit: 'In many countries a great deal of food is now being produced at the cost of undermining the existing soil resources' (K.F.S. King and Chandler, 1978).

In Andhra Pradesh, it was found that when the root systems of bajri and peanut, grown in alternate rows, were separated by vertical plastic sheets in the ground, the bajri leaves became yellow because they could not obtain nitrogen from the soil under the peanuts (Willey and Reddy, 1981). This experiment does not prove that peanuts were supplying the bajri with N, because when bajri roots were allowed to mingle with peanut roots, they might have been obtaining N from the soil beneath the peanuts while the peanuts fixed theirs from atmospheric nitrogen.

The yield of monocropped sorghum in Nigeria declined sharply after the first year that no fertilizer was added, and it continued to decline slowly every year thereafter. But when other crops were intercropped with the sorghum, this decline did not occur, and it was possible to grow profitable crops of sorghum for several years. Over time, intercropping proved 60 per cent more profitable than monocropping (Finlay, 1974).

In Côte d'Ivoire, the traditional methods of the Dranouas tribe include intercropping several varieties of yams and okra on yam hills with other intercrops in the spaces between the hills. Very similar yam-planting patterns in Jamaica probably had their origin in Africa. Manioc, cotton, pimento, maize, tomato and eggplant are grown in Côte d'Ivoire along with 15 varieties of yams – six early varieties, five late ones, and four in-between. In colonial times, when the Dranouas were forced to grow monocropped cotton as a kind of tax, very heavy soil erosion resulted. With intercropping, small farmers who do not have enough land can continue to cultivate their fields for up to nine years. If, at the end of this period of cultivation, Imperata grass comes in, the soil will not recover its fertility because the grass roots are too shallow to retrieve nutrients from deep in the soil. If the plot goes back to forest, the deep roots of the trees can obtain mineral nutrients from decaying rock deep in the soil and bring them to the surface gradually to rebuild soil fertility (FAO, 1956). The shade and competition from trees helps push back Imperata grass.

Other ways in which intercropping can rebuild soil fertility are by growing an intercrop as a green manure and by using intercrops to recover added mineral fertilizers which would be lost if the crops were not grown. In Uttar Pradesh, India, sannhemp was grown as a green manure intercropped with sugar cane. When the sannhemp was ploughed in, instead of being harvested, it increased the sugar yield by 50 per cent and also improved the soil (Bhadauria and Mathur, 1973). For farmers who are not rich enough to set aside fields for a whole season to grow a green manure crop in rotation with other crops, intercropped green manure may provide the answer.

In Maharashtra, India, the addition of P_2O_5 (phosphate) to the soil was much more effective when applied to an intercrop than to a monocrop. In one experiment, intercropped peanuts and moth bean were fertilized with phosphate, and then a bajri millet crop was grown in the same field. When the P_2O_5 was applied directly to the bajri, it increased yields by 16 per cent, but when the fertilizer was applied to the legumes which preceeded the bajri, the yield of the following bajri crop increased by 25 per cent to 30 per cent (Bhatawadekhar *et al.*, 1966). This may have been due not only to the fact that the legumes captured more of the available phosphorus, but also to the fact that they may have put the phosphorus into an organic form from which it was released slowly during the bajri-growing season instead of being applied all at once and being lost before the bajri could use most of it.

In experiments in Nigeria, intercropped maize and cassava helped build up soil fertility, whereas the addition of mineral fertilizers rapidly made the soil acid. Within three to four weeks of the application of mineral fertilizer, the soil pH dropped from 6.5 to 4.6, which increased the amount of available manganese (Mn) and aluminium (Al) in the soil. Thus, the effect of adding chemical fertilizers was to make these poisons soluble so that they could be picked up by plants (Juo and Lal, 1977). In industrial countries, acid rain seems to be speeding up this same process.

Another way in which mineral fertilizers are inferior to intercropping and organic fertilizers is that their effectiveness tends to decrease over time, partly because trace elements are used up and partly because soil structure is destroyed. In a previously unfertilized field producing 50 bushels of maize per acre, the addition of 40lbs of N gave 20 bushels additional yield. In a heavily fertilized field producing 90 bushels of maize per acre, the addition of 40lbs more of N produced only one additional bushel per acre (Terhune, 1976). On poor soil, practising intercropping can be more profitable than adding fertilizer. In the Cameroons, intercropping and traditional spacing were more profitable than row planting with or without mineral fertilizer. On richer soils, monocropping with fertilizer was more profitable (IRAT, 1978). The explanation of this phenomenon, which has been observed in many places, is that the rich soil may have only one or two elements in short supply which are limiting factors for the growth of plants. The addition of the missing element enables the crop to make use of all the other available elements; thus, dramatic results are obtained for very little expenditure. But in poor soils where most, or all, of the elements and micronutrients are scarce, the addition of one or two elements through the use of mineral fertilizer will have little effect. This is one reason why intercropping is important for farmers with poor soils and why it will be important for all farmers in the future. The system helps to recycle all the nutrients in the soil and helps to prevent their loss through leaching or erosion. Animal manure, compost, mulch, sewage and intercropping with a green manure crop will all be important sources of nutrients in the agriculture of the future.

Labour

Figs 21 and 22 show that two-crop intercropping in Nigeria is more profitable than monocropping, that three crops grown together are more profitable than one, and that four-crop intercropping is more profitable still, whether or not outside labour must be hired.

Fig. 21 shows costs calculated as if the family members got paid for their work. When fields are ploughed, cultivated or weeded by hand, it is not much more work to raise two crops at once or even three or four. The net profit goes from about 80 shillings/acre (sh/ac) for one crop up to almost 200sh/ac for four crops grown at once. The attitude of the commercial farmer is brought out clearly by these graphs. The farmer who is interested in a monetary profit would rather grow two crops than three crops, while the family farmer would prefer to

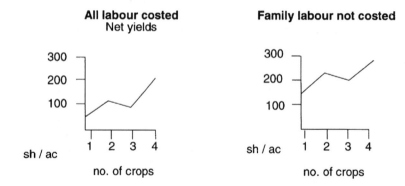

Figs 21, 22: Efficiency of labour use (value of yield in shillings per acre) (Nigeria) Source: Norman, 1977.

grow three than two. In the case of the commercial farmer, if the cost of labour is more than the value of the crop, he will not grow the crop. For the family farmer who has some family labour available for whom there is no other source of employment, it makes sense to grow the extra crop since it will provide food or cash for people who would not otherwise have employment, food or clothing.

Fig. 22 shows how profitable it would be to intercrop with several crops at once, if family labour is regarded as being free or as working for room and board. The strict economist or commercial farmer may calculate that the effective wage per hour of family labour is very low and then argue that these people are being exploited. The commercial farmer may pay better wages than this, but he hires people only when he needs help. He pays no wage income when he does not need help, he hires no one when the cost of labour is equal to, or more than, the value of an intercrop, and he will introduce labour-saving machinery and grow only one crop per field regardless of social and ecological consequences if that is the most profitable method.

This example makes clear that the difference between commercial and traditional farming is the same as the difference between farming for maximum profit and farming for maximum yield (Norman, 1977). The large-scale farmer who is interested in maximizing profits will tend to grow only the most profitable crop. Since small amounts of secondary and tertiary crops might be expensive to process, market and transport, these extra crops tend not to be grown on big farms; and since landless labourers have little money, it may not pay the landowner to intercrop his main cash crop with food crops for sale, especially if the second or third crop tends to lower the yield of the main crop. These factors explain why commercial plantations in the tropics are usually monocropped (Norman, 1977).

When hand labour is being used to produce one crop, very little extra labour is needed to intercrop (D.W. Norman (1974b); Finlay, 1974; Fisher 1977). On a Taiwanese family farm, intercropping involved practically no extra cash outflow while increasing yields by 100 per cent (Beets, 1975). Since the labour needed to grow coffee in Kerala, India, involves digging, trenching, manuring, weeding and irrigating, no extra labour is needed for intercrops (Chengappa and Rebello,

1977). In Tamil Nadu, as in much of monsoon India, it saves labour and time to plant several crops at once by mixing seeds in the seed drill, and it helps solve labour problems at harvest time because the harvesting of one species after another spreads the work out over a longer period. In this way, a family can do the work and not have to hire labourers or harvesting machinery (Ayyengar and Ayyar, 1941). Whereas a big farmer wants all the harvest to be ready on one day so that it can be done with a machine, the small farmer wants the harvest spread out over time so that the work to be done is compatible with the available labour (Kaushik, 1951; Jodha, 1977). A few of the many intercropped combinations which accomplish this are maize and beans (T. Anderson, 1976; Hasselbach and Ndegwa, 1980) and cassava in intercropped mixtures (Porto *et al.*, 1979). The following chapters based on field observations in Jamaica, Nepal and India give many more impressive examples.

Some intercropping combinations are so profitable that it pays even big farmers to plant them. In Karnataka, India, cotton and onions turned out to be a profitable and labour-intensive mixture. Compared to monocropped cotton, the two crops together used 166 per cent more men (112 men/acre) and 133 per cent more oxen (28 bullocks); in spite of this, the net monetary increase was 24 per cent (Kubsad and Dasaraddi, 1974). But the usual practice is for big farms not to intercrop. In Malaysia, most of the area devoted to rubber is monocropped, but it is quite easy for a family that owns a plot of rubber trees to feed itself from the crops grown between the trees. Even more remarkable is the fact that the rubber yield/hectare is higher on the family farms than on the plantations. The big rubber farmer in Malaysia tends to have fewer and bigger trees/hectare than the small farmer to reduce the labour cost of tapping. The small farmer has more and smaller trees and does more work, but he also gets more rubber and more income per hectare (Bangham, 1946; Williams, 1975).

As for the agriculture of the future, it is important to realize that a buffalo expends 800 kilocalories of energy/hour of work, while a tractor expends 90,000 kilocalories (Terhune, 1976). Even though a tractor takes two hours to do the work that a pair of buffalo would do in two days, it uses seven times as much energy to plough the same area. In the future, as energy prices rise, fewer and fewer farmers will be able to afford tractors, and world agriculture will be in a position to advance to forms of motive power which do not pollute the air or use non-renewable resources.

Insects

Fig. 23 depicts the results of an experiment in the Netherlands to determine the effect of intercropping cabbage with spurry (*Spergula arvensis*), a small forage or green manure crop.

The average caterpillar population per 100 cabbage plants, when there was no intercropping, was 118 (equivalent to 100 per cent in Fig. 23). This is the average for four species of insects, all of which were reduced in number by intercropping. When there was only one spurry plant per eight cabbage plants, there were only 77 per cent as many insects, while one spurry plant per two cabbages reduced the percentage of insects to 40.5 per cent of the insects found in monocropped cabbage, where all other factors were equal (Theunissen and Den Onden, 1980).

Figure 24 shows how the number of insects was reduced by intercropping in a

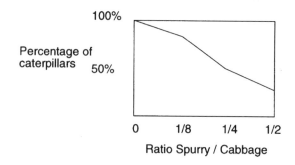

Fig. 23: Insects reduced by intercropping (Netherlands)
Source: Theunissen and Den Ouden, 1980.

Costa Rican experiment. Crop combinations which produced the most shade had the fewest insects. Such crop combinations are more effective than others at reducing insect population, which presumably reduces insect damage and the incidence of those plant diseases which are carried by insects (Moreno, n.d.).

Not all intercropping combinations will reduce insect and disease damage because some insects eat more than one crop (Willey, 1979). It was found in India that a pest named Pyrilla increased in numbers when the two grains jowar and bajri were grown together, but its population was reduced considerably when jowar was intercropped with the legumes, pigeon pea or blackgram (De and Singh, 1979). Similarly, the number of Jassid beetles on blackgram increased 20 per cent when it was grown with greengram but decreased 25 per cent when it was intercropped with jowar and 50 per cent with bajri (K.M. Singh and R.M. Singh, 1977).

Another example of reduction of insect pests by intercropping is Swedish forests, which are thought to have fewer outbreaks of insects than forests in Germany because there is a greater mixture of tree species in Sweden. The big ICRISAT crop research centre in Hyderabad, India, suffers more insect damage than neighbouring small farmers who intercrop. ICRISAT is responsible for keeping alive the seeds of many varieties of, for example, pigeon pea and sorghum, and the corresponding insects are attracted to these large, monocropped, seed-reproducing areas (Bhatnagar and Davies, 1979).

In many cases, insecticides are making insect problems worse. In Britain, the spraying of orchards to kill various insects has made the immune fruit-tree red spider mite into a much more common pest because its natural enemies are killed by the spray (Dempster and Coaker, 1972). In Colombia, planting cassava and beans at varying populations of plants per acre showed that intercropping always gave higher yields than monocropping. The use of insecticides tended to be less effective than the practice of intercropping in controlling pests. In two-plant density experiments, intercropping was as effective as insecticides in reducing the number of insects, while in three other experiments the intercropped plots with no insecticides had 30 per cent fewer insects than did monocropped plots with insecticides (Thung and Cock, 1979). Some 240 insect species have now developed resistance to insecticides. It is only recently that the importance of

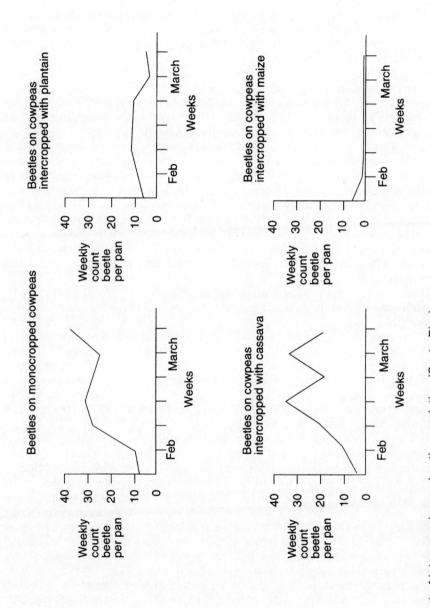

Fig. 24: Effect of intercropping on beetle population (Costa Rica)
Source: Moreno, n.d.

natural pest control has been appreciated. When spray kills the predator insects that formerly kept the plant-eating insects under control, crop damage from insects often increases. One reason for this new problem is that there are many more plant-eating insects than predators (just as there are more rabbits than wolves), thus the plant-damaging insects are more likely, statistically, to develop immune mutations (Rodale, 1983). Researchers are now doing experiments on intercropping as a way of controlling insects. For small farmers practising intercropping, insects are seldom a great problem, partly for the reasons mentioned below and partly because small farmers can go through their fields frequently, picking off and destroying insects before they can become numerous enough to do significant damage.

There are several ways in which intercropping can control pests. One theory is that aromatic or strong-smelling intercrops will keep away pests. This may be true but has not been proven scientifically. Chemical repulsion may be one operative factor, but physical controls and predator havens, discussed below, may be other controls operating at the same time. Some researchers believe that chemical controls are involved in cases where *Aphis brassica* and other pests of mustard and taramira are kept away by the chickpea crop, and similarly that chickpea pests are repelled by mustard and taramira (Kaushik, 1951). This plant protection increased yields by 200–400kg/ha. In Canada, carrot rust fly may be averted if onion, garlic, leeks or chives are used singly or in combination as carrot intercrops (Stewart, 1980). Stewart also says that buckwheat intercropped with beans will attract hover flies which prey on the aphids that affect beans, and that tomatoes and asparagus are a good combination because they protect tomatoes from a species of nematode and asparagus from the asparagus beetle. As mentioned in Chapter 4, asparagus and tomato are a favourite combination in Kashmir, India.

But experiments at Emmaus, Pennsylvania, found no long-term insect control was effected by aromatic plants, and when they were intercropped with broccoli, all of the experimental aromatic plants attracted extra insects to the vegetable (Matthews, 1981). The truth of the matter is not yet known. Aromatic plants may work in some places but not in others. The author's fieldwork, discussed in later chapters, found little dependence by Third World farmers on intercrops which were aromatic but had no other use.

The physical confusion of insects by intercropping is more generally accepted by researchers as an explanation for its success in controlling insect pests. In a monocropped field, the insects which eat the crop have easy success in finding crops to eat and suitable places to lay eggs. In northern Nigeria, a pest called Maruca causes great damage in commercial fields of monocropped cowpeas. The traditional method of growing cowpeas is as an intercrop in sorghum–millet fields. The grains shade the cowpeas, thus making the plants smaller, and these small cowpeas do not attract Maruca (Baker and Yusuf, n.d.).

In Costa Rica, experiments with three crops grown together reduced the presence of insects significantly in 56 per cent of 310 observations: maize and beans together reduced insects in 39 per cent of the observations, maize and squash in 49 per cent (Risch, 1980). This is one of the few experiments which tries to replicate indigenous methods. The closer one comes to traditional intercropping with several compatible crops grown at once, the greater the benefits become. With regard to individual beetles, Risch found that *A. Thiemi* was on squash plants three to ten times less in intercropped than in monocropped plots. Light seemed to be a factor in attracting beetles. The more shade created by intercropping in the experimental plots, the fewer beetles were found, and the more likely they were to fly out of the plot. *D. Balteata*, which also eats squash, occurred two to ten times less frequently in intercropped plots. Bean-eating beetles were more hindered by maize than by squash. Some beetles ate two of the three crops and became more numerous when those two crops were grown together. But the presence of even one crop which a particular beetle did not eat always lowered the number of beetles (Adesiyun, 1979; Ramdhawa, 1975; Risch, 1980; K.M. and R.N. Singh, 1977). It is not really surprising that intercropping is so effective in controlling insects. Not only did pre-literate people do an enormous amount of work to develop varieties and combinations which could survive, but the wild plants from which domestic plants are descended had previously evolved diversities, immunities and combinations in the wild which made it possible to survive insect attack.

Intercrops provide a home base for useful predator insects and spiders that attack bugs which damage the main crop. This seems to be an important method of control and has no bad effects on the men, women and children who work in the fields. Harmful insects cannot become immune to this method of control.

Heliothus bugs, which eat cotton in Peru, can be almost completely eliminated by planting maize every tenth row. Four species of useful bugs hatch out in the silk of the maize, and the adults fly to the cotton where they eat the eggs and young of *Heliothus*. Some *Heliothus* lay their eggs on the maize, where they are even more easily disposed of by predators. For the greatest success of this method, a variety of maize should be chosen which produces silk at the appropriate time to attract *Heliothus* egg-laying and to nurture a sufficient number of predators to attack *Heliothus* at the egg and larvae stage (Simon, 1954). Two species of leaf-roller which attack cotton were also controlled by growing maize nearby. The plots of maize enabled predators to survive from one cotton season to the next. Plots of maize near cotton increased the number of spiders by 70 per cent; intercropping in alternate rows produced 88 per cent more spiders than monocropping. Intercropping with maize increased the numbers of all beneficial species by 90 per cent. In the Philippines, tomatoes intercropped with cabbage reduced attacks of the diamond-back moth and helped delay or prevent the introduction of increasingly dangerous insecticides. For similar-sized plots, monocropped cabbage had 127 moths and 2 507 eggs; cabbage intercropped with tomato had 43 moths and 1867 eggs (Buranday, 1975).

When using intercropping to control insects, it is important not to choose crops which will increase insect problems. In northern Nigeria, for example, it happened that the timing of maize and tomato crops produced large numbers of two insect species which were ready to attack cotton in nearby fields when the cotton reached a vulnerable stage. Growing maize or tomato species with different dates of maturity should help solve this problem (Bhatnagar and Davies, 1979). The general effect of traditional intercropping, using species and varieties developed by local people, is to have high populations of predators ready when they are needed. The multiple intercropping patterns of indigenous farmers is much more effective overall in controlling insects than the vulnerable monocropping practice of modern farming. Many modern high-yield crop varieties were bred with the assumption that insecticides would always be used, but as the effectiveness of insecticides decreases, the method of controlling insects by intercropping becomes more important. Complex ecosystems have fewer drastic swings in insect population than simple systems. In

Costa Rica, squash blossoms in intercropped plots provided food for parasitic wasps (Risch, 1980). In India, the small gourds which are often grown in intercropped grain fields may be more valuable for encouraging insect predators than for producing their small, wizened fruits.

In southern Uganda, the prolonged presence of vegetation in intercropped fields makes it possible for predators to survive from one season to the next and keep *Heliothus* under control. In southern Tanzania, where the dry season is longer, less vegetation survives. So there are fewer predators in the drier area, and *Heliothus* is able to attack cotton relatively unchecked. The higher humidity of dense intercropped foliage benefits fungi that attack insects. This is thought to explain the relative lack of mites in areca plantations which are intercropped with banana (Khader and Anthony, 1968; Bhatnagar and Davies, 1979).

Trap crops are another intercropping method for controlling pests. Brussels sprouts and sugar beet in Britain, for example, had far more damaging insects in plots where weeds were removed than in plots where weeds remained. This was because weeding deprived insects of an alternative food source. In an experiment where clover was planted as an intercrop (a substitute for the weeds), there were many fewer damaging insects on Brussels sprout leaves and roots. The clover made it harder for insects to find the sprouts, and it also acted as a home base for predators (Dempster and Coaker, 1972). Alfalfa has been found to be a good trap crop for cotton insects (Stern, 1961). An experiment in Israel showed that growing two varieties of sugar beet lessened the attack of birds on the seedling beets. With no extra beets as a bait crop, only 3.6 per cent of the seedlings survived for three weeks, but with the beet-bait crop, 19.5 per cent of the seedlings survived the skylark menace (Benjamini, 1980).

Prior to modern breeding practices, seeds were selected from plants that had survived best under native practices. This has many practical good effects, among which may have been selection of alleles resistant to pests (Pimentel, D. and M., 1977; Bhatnagar and Davies, 1979). A European who noticed the lack of concern of a Sierra Leone farmer about caterpillars attacking his young orange trees, wondered if the farmer was in this way selecting oranges most resistant to insect attack (Ehrenfeld, 1978).

When all else fails, the indigenous intercropper has one last line of defence against insect attack. If insects severely damage one crop, more sunlight can reach another crop and somewhat make up for the damage. In an experiment in the Philippines, when maize leaves were cut off to simulate insect damage, intercropped peanuts yielded 50 per cent more. If one-third of the maize plants were pulled up 25 days after planting, the total yield was enlarged because the remaining plants had more environmental resources available and grew enough to make up for the missing plants. When simulated damage occurred late in the season (70 days after planting), the losses were not completely made up. In monocropping situations, there was less recovery from simulated insect damage (Liboon *et al.*, 1976).

Disease

Control of insects helps control disease since insects often carry disease from one plant to the next. In Costa Rica, the incidence of powdery mildew and cassava scab on cassava was less when the crop was intercropped with beans and sweet potato but more when it was intercropped with maize. Cassava rust was 60 per cent less severe when there was intercropping with maize plus beans, 35 per cent less with maize, and 42 per cent less with beans. Cassava dieback disease occurred 50 per cent less with multiple intercropping of maize and beans (Moreno, n.d.). Cowpea mosaic disease also spread less when beetle activity was restricted by intercropping. The two beetles which spread this disease were less active when more shade was created by intercropping cassava with plantain. The importance of shade in this case is shown by the fact that 100 per cent of the cowpeas were infected 45 days after planting when intercropped with maize. Cowpea infection was only 19 per cent and 18 per cent, respectively, 66 days after planting when intercropped with cassava and plantain (Moreno, n.d.) (see Fig. 24).

The fact that there is less disease in intercropped fields is quite well established. Figs 25 and 26 show the results of a Tanzanian experiment.

Fungicide did not control powdery mildew on greengram quite as well as intercropping did (Keswani and Mreta, 1980). Intercropping is, of course, less dangerous than fungicide, less expensive, and provides more food and employment.

Sorghum had fewer pests and less disease in Tanzania when it was intercropped with peanut, cowpea or sweet potato (Kayumbo and Asman, 1980). In Kenya, also, intercropping gave protection from pests and disease when chemicals were not used (Fisher, 1977). In Uganda, intercropped beans lowered the incidence of peanut rosette disease considerably (Mukiibi, n.d.). In Nigeria, monocropped cowpeas were more subject to disease than when intercropped with pearl millet (Norman, 1974). In Costa Rica, as described in the discussion of insects and intercropping, cassava with maize and/or beans reduced the severity of several diseases (Moreno, n.d.). In Venezuela, indigenous intercropping was observed to reduce pest and disease attacks (Sebastiani, 1981), while in Sweden, it was found that fewer pea seeds were infected with fungus when peas were intercropped with oats (Bengtsson, 1973). However, as in the case of insects, unwise intercropping combinations or timing can sometimes increase the problem instead of reducing it (Willey, 1979).

In monocropped fields, where all the crops are the same species and even the same variety, plants of the same species touch each other, allowing disease to spread easily. Since some diseases are becoming resistant to spray, disease control by interspersing plants of an immune species throughout the field will be important in the future. When rust spores land on the wrong plants no disease occurs, and no additional rust spores are produced to spread like a chain reaction throughout the field (Johnson and Allen, 1975). This will result in a much healthier field than a monocropped one in which rust spores which land on leaves are almost always successful in infecting the plant and in producing more spores. The complete failure of monocropped rubber plantations in Brazil during World War II is one of the best known examples of this effect.

When indigenous intercroppers choose the seeds of plants which survive as seed stock for next year's planting, they automatically select for disease resistance. Plant breeding on experimental farms, where insecticides and fungicides are applied as standard practice, tends not to produce disease-resistant varieties. Use of chemical insecticides and fungicides is, of course, more costly than disease control by intercropping. Spraying intercropped fields is difficult and wasteful, not only because it is hard to get the right spray on the right crop but

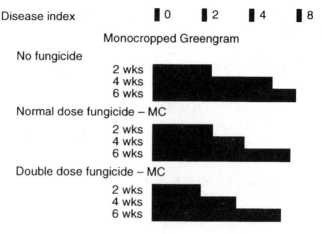

Fig. 25: Disease control by fungicide in monocrops (Tanzania)
Source: Keswani and Mreta, 1980.

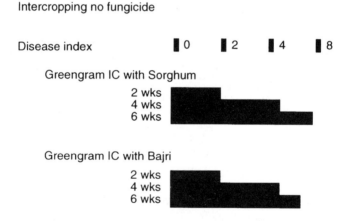

Fig. 26: Disease control by intercropping (Tanzania)
Source: Keswani and Mreta, 1980.

because spray may damage intercrops. Banana fungicide sprayed from the air in Jamaica damaged the small farmers' intercropped coffee. This practice has now been replaced by ground spraying and by planting disease-resistant bananas. In Taiwan, disease-resistant maize was helpful in making maize-sugar cane intercropping combinations possible. Previously, both sugar and maize suffered from the same downy mildew (Cheng, 1979).

In southern California, a trap crop to prevent the spread of an aphid-borne disease which discoloured melons was developed. American consumers will reject muskmelons or canteloupes which have a harmless fungus discolouration on the surface. Fungicides and insecticides to control this discolouration are far from harmless; experiments were conducted, therefore, to find a crop which the disease-bearing aphids would prefer to melons. Radish and Swiss chard attracted aphids and provided good protection from disease but shaded the melons so much that they could not ripen. Wheat turned out to be an ideal trap crop. A plot of canteloupes with no protection was 82 per cent mottled by virus a week before harvest. Spraying the fruits and leaves with a light oil reduced the mottling to 40 per cent of the fruit, while planting adjacent wheat reduced mottling to 4 per cent (Toba *et al.*, 1977).

Weeds
Weeding is necessary only in the first few weeks with traditional agriculture because the additional crops are soon able to suppress the weeds (Waddell, 1972). Weeds taken from these fields are usually carried home to feed livestock and supplement their diet of stalks and leaves from grain crops which have been harvested. The integration of animals into traditional farming systems, in order to make use of biomass which humans cannot digest, is a very important aspect of traditional methods.

Weeds become a very serious problem wherever monocropping is adopted. The inter-row spaces provide room for weeds to flourish (Gahlot, 1978). In the United Kingdom, it has been calculated that each ten-fold increase in weed density lowers production of the main crop by 25 per cent (Haizel, 1972). Since weeds sometimes reduce the yield of the main crop as much as an intercrop would, it would seem to make sense to encourage intercropping instead of constant weeding (Moody and Shetty, 1979), thus producing more food for less work. If weeding is done for the first 30 days, the weeds will not be able to damage the main crop very much thereafter (Haizel, 1972). This suggests the importance of fast-growing cover crops to utilize the sunlight in spaces between plants of a slower-growing main crop and to deny sunlight, nutrients, water and space to weeds. In Malaysia, it was found that long-bean, ginger, okra, tomatoes, cucumber and peanut were all good intercrops for small farmers to grow with pineapple. Pineapple takes 16 to 18 months from planting to maturity; thus, intercrops not only suppress weeds but give good extra income during the period before the pineapples can be sold (Lee, 1972). A monocropped field of maize produced 4 tons/ha of weeds 40 days after sowing, but maize intercropped with soybean produced only 0.5 tons/ha of weeds (Moody and Shetty, 1979).

Using herbicides to try to solve the weed problems associated with monocropping has led to massive soil erosion from bare inter-row and post-harvest spaces; another problem is that weeds, like insects, can become immune to poisons. Some rhizomatous weeds are now resistant to paraquat and persist in no-till fields (Okigbo and Lal, 1977). In the Philippines, a herbicide-resistant sedge (*S. maritimus*) became a serious problem when rice was monocropped in the same field for eight years. Crop rotation with maize and peanuts intercropped helped to control this infestation (Lacsina and De Datta, 1975).

Herbicides are difficult to use in intercropped fields because of the damage they may do to one of the crops. There are a few two-crop combinations for which specific herbicides will damage neither crop, but such spraying is a complicated procedure. In Taiwan, a herbicide was found which allowed sugar to be intercropped with maize or sweet potato, but costs are high, application must be precise, and the number of intercrop combinations is limited. The more tolerant chemicals are of crops, the more tolerant they are of weeds (Moody and

Shetty, 1979). In Georgia, USA, it was difficult to intercrop maize and soybeans, or sorghum and soybeans, because most of the usual weed-killing chemicals would kill the soybeans (Cummins, 1973). But when intercropping is used to control weeds, chemical herbicides are not needed.

A well chosen, quick-growing intercrop can suppress weeds in a field of a long-term crop quite effectively. In northern Nigeria, early millet suppresses weeds in sorghum; in the humid tropics, cucurbits or cowpeas suppress weeds among slow-growing root crops; and in East Africa, beans suppress weeds in maize (Fisher, 1977). In Tanzania, legumes such as peanuts and soybeans, and grains such as millet, maize, or sorghum were grown together to control weeds. Legumes alone were soon overwhelmed with weeds; monocropped soybeans yielded 97 per cent less when weeds were not controlled. Monocropped pearl millet gave 57 per cent less yield when there was no weed control, but soybeans and millet grown together had only a 46 per cent reduction in yield where weeds were not controlled (Mugabe et al., 1980).

In India, greengram was effective in controlling weeds between rows of sugar cane (Sivaraman, 1973). In Costa Rica, monocropped rubber was infiltrated by grasses and other weeds, but when intercropped with cacao, the ground was so shaded that few weeds grew (Hunter and Camacho, 1961). In Nigeria, less weeding was needed when plantain and cocoyam were intercropped than when plantain was grown alone (Devos and Wilson, 1979). In Costa Rica, the crop's proportion of the total biomass (including weeds) in monocropped plots of maize, cassava and beans was 20 per cent, 25 per cent, and 85 per cent, respectively; when the three were intercropped, they made up 84 per cent of the plant biomass, and weeds only 16 per cent (Moreno, n.d. quoting Hart). In another experiment, mung beans grew more quickly and suppressed weeds better than peanut or sweet potato in intercropped maize. In coconut plantations, maize plus mung beans eliminated the need for weeding. Some experiments in agricultural economics found that weeding intercropped fields is more expensive than weeding monocropped fields, but other researchers have found the reverse to be true (Moody and Shetty, 1979).

At ICRISAT in Hyderabad, density of planting was varied to discover the effect on weeds and on yields. In Fig. 27, for the two monocrops, the dry matter weight of weeds at the time of sorghum harvest was $30 gm/m^2$ for monocropped jowar and $169 gm/m^2$ for monocropped pigeon pea.

The data in Fig. 27 show clearly how weeds are controlled by increasing the density of useful crops. Intercropping helps control weeds, at the same time giving higher yields.

Erosion

Intercropping controls soil erosion by preventing raindrops from hitting bare soil where they tend to seal surface pores, prevent water from entering the soil, and increase surface erosion. Extra root systems hold on to soil, and a higher organic content in the soil results from the decay of extra crops. Organic matter holds water in the soil and also increases the number of soil fauna which make holes in the soil, allowing water to enter. Dead mulch on the soil surface retains water, as does live mulch, which can be an intercrop. The mulch crop may die without being harvested or may be ploughed into the soil (Finlay, 1974; Greenland, 1975; Kaushik, 1951; Lal, 1975; Okigbo and Lal, 1977).

Indigenous farmers who intercrop have developed many successful crop com-

Population	Weeds in IC fields as % of weeds in MC fields (by wt)	LER
½ NS and ½ NPP	36	0.99
½ NS and NPP	32	1.17
½ NS and 2 NPP	25	1.22
NS and ½ NPP	23	1.19
NS and NPP	15	1.27
NS and 2 NPP	18	1.53
2 NS and ½ NPP	10	1.05
2 NS and NPP	10	1.54
2 NS and 2 NPP	9	1.40

Fig. 27: Effect of intercropping and crop density on weeds and yields (India)
Notes: N = Normal population density (180 000 plants/ha for sorghum, 40 000 plants/ha for pigeon peas). S = sorghum, PP = pigeon peas.
Source: Shetty and Rao, 1979.

binations to control erosion, as the following chapters suggest. Experiments, even when only two crops are involved, have shown the superiority of intercropping as a control mechanism. In Illinois, rye broadcast in maize during August or September grew to sufficient size to hold the soil during the winter. Legumes also held the soil in maize fields after the maize harvest, but did not form as thick a sod as rye (Kurtz, 1952). In East Africa, beans in maize fields reduced soil loss, just as cowpeas did in West African sorghum fields (Fisher, 1977). In Gujerat, India, a peanut cover crop in cotton fields allowed only 50 per cent of the soil losses from cottom fields (Joshi and Joshi, 1965). In Himachal Pradesh, India, a dense canopy of soybean-intercropped maize can reduce soil losses even on sloping land (C.M. Singh and Chand, 1980).

The largely bare soil of modern monocropped fields has allowed erosion in America to proceed two, three or more times faster than the rate of soil formation. To increase profits by neglecting erosion control ensures that profits will exist only in the short term. Future farmers will find themselves working under very difficult conditions because of the soil-destructive practices of the present (*Global 2000 Report*; Sampson, 1981).

2 Intercropping in the Christiana area of Jamaica

Historical background

AFTER DISCUSSING THE HISTORICAL background of Jamaica's traditional agriculture, this chapter records the intercropping combinations used in an area of the island where many traditional techniques have survived. After recording the frequency with which different intercropping combinations are used and showing how maps can record the proportion of each crop in a combination, the chapter concludes with various methods of analysing the data and a study of the effect of intercropping on soil fertility.

The difference between the attitude towards agriculture of modern scientifically-backed, profit-oriented monocropping and traditional, intuitive, multi-purpose intercropping is so great that it is necessary here to attempt an initial statement to explain the point of view of the indigenous small farmer. Many readers may ask how it is possible that poor people with little formal education could practise a type of agriculture which is, in many ways, superior to modern, scientific agriculture. The short answer to this question is that any culture groups which used agricultural methods which destroyed the soil have not survived.

Since Jamaica is, in effect, an African island as far as small farmers are concerned, two reminders of other African achievements might be helpful. Viewing the Miró and Picasso paintings at the Museum of Modern Art in New York City, followed immediately by a visit to the African art section in the Michael Rockefeller gallery of the Metropolitan Museum, impresses one with the inventiveness, humour and sophistication of African artists. Likewise, in American popular music, there is a serious African influence. Listening to recordings of indigenous music played in Africa will almost certainly show the listener the superior sophistication and inventiveness of undiluted African music. In this chapter, a description of traditional agriculture in Central Jamaica will show that it is far more complex and achieves many more goals than modern mechanized agriculture. The great complexity of one of the fields recorded and analysed in this chapter is suggested by Fig. 39 (see p. 45).

The way in which Europeans became aware of African culture, and most world culture, was not conducive to its understanding and appreciation. Christian missionaries never doubted that their monotheism was superior in every way to polytheistic religions. Modern scientists seem to have inherited the attitudes of the missionaries. Whereas religion is a matter of faith, the ideas of scientists, as we have seen in Chapter 1, can be tested objectively. For many years Europeans claimed that agriculture of European origin was superior to indigenous agriculture without ever testing the efficiency with which indigenous agriculture uses resources, conserves soil and provides employment. Recent research on indigenous methods (see Chapter 1) suggests that world hunger and soil destruction result at least partly from the abandonment of traditional agriculture and intercropping (Richards et al., 1989).

Since Jamaica is a large island, there was room for many of the slaves who were brought from Africa to grow their own food; thus, African intercropping skills were preserved in Jamaica. Little garden-type farms came to supply much of the island's food; these intercropped fields could supply modern food needs if small farmers had more land. Small farmers not only produce food for themselves but can grow the same tropical export crops that are currently grown mainly on big farms. In 1867 it was small Jamaican farmers using intercropping who supplied the bananas which began the banana export trade. Big farmers have found it more profitable in the short run to grow only bananas and to use no intercropping. The subsequent infection of the soil with Panama disease may be encouraged by monocropping, but if profit goals are short-term enough, the fruit company may still be economically successful before the land is abandoned. Perhaps modern economics, like narrowly-focused science, is to a certain degree responsible for world food shortages.

Field observations

This study of small-farm agriculture was carried out both east of Christiana in non-limestone country and west of the village in the limestone area. The author talked to farmers and collected field data by writing down the intercropping combinations which were observed in each field. In the 1951 and 1955 field seasons, most data were not recorded by main crop or by geologic area; these preliminary data were used to construct Fig. 29 (see p. 36). The bulk of the data for the remainder of this study are from the 1961 and 1971 field seasons. The 1981 field season was devoted to checking information and ideas without collecting more data on intercropping combinations.

Instead of attempting a random sample of fields, every field seen was recorded. All the paths and roads in the Christiana area were traversed in an attempt to get a sufficiently large database. Trees were recorded if they grew only once in a field, but even in small fields other crops had to occur at least three times to be included. Fields contain volunteer crops, such as taro, but such is the intensity of supervision and cultivation here that it is usually accurate to assume that decisions have been made about allowing individual volunteers to remain; they can, therefore, be counted as crops. Instead of including scientific names in the text, most of this information is included in Appendix 2. For more information about descriptions and uses of these crops, the standard reference book used is Purseglove's *Tropical Crops*.

Since the total number of fields for which main crops were determined is 998, the percentage occurrence of each main crop can be calculated by moving the decimal place. It is important to realize that Fig. 28 does not include all the crops grown by small farmers since it does not include the intercrops.

Fig. 29 lists the number of intercrops found with ten main crops. These ten crops were major crops in 956 fields, but when their occurrence as secondary crops is also included, they were found in 3758 fields. Thus, it would almost be true to say that for every field where a crop was a major crop, there were three other fields in which it appeared as a secondary crop. The small farmer would be very much poorer if he or she did not intercrop.

Some crops are easier to grow as intercrops than others. If a crop is important as a cash crop, then the farmer will cater to its peculiarities and add only secondary crops which will not hinder it too much. This is true of Irish potato. Christiana, with an elevation of around 2000 feet, is one of the cooler parts of Jamaica and the country's main supply area for Irish potatoes. This explains why Irish potatoes are the main crop in 77 per cent of the fields where they are grown. Taro, at the opposite end of the scale, is very easy to grow as an intercrop since some varieties do well in shade while others do well in full sunlight. Taro is sold

Banana	346	Coffee	14	Pigeon Pea	3
Cabbage	40	Ginger	11	Pumpkin	5
Carrot	2	(Irish) potato	139	Scallion	4
Cassava	4	Kidney bean	47	Sugar	5
Chilli	1	Lima bean	1	Sweet potato	71
Chocho (chayote)	16	Maize	99	Taro	2
Citrus	17	Peanuts	3	Tomato	6
				Yam	182

Fig. 28: The number of fields in which farmers have chosen these crops to be the main crop (Christiana, Jamaica)

No. of crops per field	1	2	3	4	5	6	7	8	9	A	B	C
Percentage of fields with 1,2,3 crops, etc.	6	21	25	20	12	7	5	2	2	956	3 758	–
Main crop												
Banana	2	19	26	21	12	10	4	3	3	346	616	56
Cabbage	19	33	24	14	5	–	–	5	–	40	80	50
Chocho	20	15	5	10	5	10	5	5	15	16	91	18
Coffee	–	17	36	19	11	7	2	2	5	14	321	4
Irish potato	4	12	15	25	23	6	12	2	–	139	181	77
Kidney Bean	6	21	15	21	15	4	15	2	–	47	227	21
Maize	12	21	22	21	11	6	5	2	1	99	514	19
Sweet potato	10	32	24	15	9	6	2	2	1	71	276	26
Taro	1	17	28	25	14	7	5	2	1	2	77	0.2
Yam	7	22	25	20	12	6	4	2	1	182	665	27

Fig. 29: percentage intercropping in fields showing how often crop combinations had one, two, three crops, etc. (Christiana, Jamaica)
Note: Columns 1–9 show the percentage of fields with 1 crop, 2 crops, 3 crops, etc., up to 9 crops; column A gives the number of fields in which each listed crop was the main crop; column B gives the total number of fields in which each listed crop was found; column C shows the percentage of fields where the crop was grown in which it was the main crop.

as a cash crop but only to poor people who cannot make demands about its size and appearance. Since taro comes up year after year from the same roots if fields are cultivated, some of its frequency as an intercrop can be explained by its enthusiasm as a volunteer.

It can be seen from columns 1 to 9 of Fig. 29 that half the fields have four crops or more. When it is remembered, from Chapter 1, that almost all modern intercropping research is done with only two crops per field, it will be understood that modern agricultural research is still far from an objective appraisal of an indigenous farmer's agricultural efficiency. The vertical black line in Fig. 29 divides the total percentage for the ten crops almost in half with 52 per cent of the fields having three crops or less and 48 per cent having four crops or more. About half the banana, coffee, corn, taro and yam fields have three crops or less and the other half have four crops or more. Cabbage is not often intercropped because it is a commercial crop. Cabbage and other vegetables and fruits sold commercially often sell better if the individual item for sale is larger rather than small. Whether customers naturally prefer bigger fruits and vegetables, or whether they have been educated to prefer them by modern monocropping and salesmanship, is an interesting question. The social implications of monocropping for big, beautiful fruits is not an idle question in Jamaica. Small farmers in modern Jamaica are generally unable to sell bananas for export because of the size and appearance standards which have been set.

Sweet potatoes tend to be grown alone, or in less intercropped fields, because they are the last crop to be planted in the crop rotation sequence which begins by clearing a field and ends by allowing it to go back to pasture or brush. Sweet potatoes are often interplanted in yam fields well before yam harvest and tend to spread their vines over the whole field during the months of yam harvest. The sweet potatoes hold the soil and provide food for many additional months before they are gradually replaced by weeds, brush and grass.

Chocho and kidney beans are intercropped more than the average (see Fig. 29). Chocho, the Jamaican name for chayote (see Appendix 2 for Latin names) is a vine which can climb over other plants or trees. It is often grown on horizontal trellises above the ground and can be intercroped when shade-loving plants or seedlings are grown underneath the trellis. Kidney beans are quite a small plant, probably the smallest on this list of ten crops, and are the only legume on the list. They are usually planted in spaces between large, slower-growing plants, but whole fields devoted to kidney beans can also be found. 'Main crop', or 'the crop', are classifications used by small farmers who think of the secondary crops as being 'dropped in' between plants of the major crop. Jamaican house gardens have no main crops and are not discussed here.

Bananas (in the limestone area)

For Jamaican small farmers, bananas and plantains are important food crops which lend themselves well to intercropping, as Fig. 30 suggests.

Bananas were a cash export crop for small farmers until special methods of treatment and new standards were established which tend to exclude the small farmer. Coffee bushes in small-farm Jamaica need shade; since the bushes live for 20 or 30 years, a constant succession of annual banana crops is needed. Coffee from small Jamaican farms is still an export crop and is grown in 55 per cent of banana fields. In addition to the 13 intercrops with banana shown here, 16 other crops were recorded in at least one banana field. A small farmer without much land must plan where he can plant needed crops in fields which are already devoted to some main crop.

In addition to 29 smaller intercrops, there are at least 14 kinds of trees in Jamaican banana fields in addition to the five trees shown in Fig. 30. Since a field with coffee and shade plants is already devoted to tall crops, banana fields are an excellent place for a small farmer to grow trees, as long as these are not grown so thickly that they shade the bananas unduly.

In Fig. 31 the effect of one banana intercrop on other intercrops is examined. The first eight columns show in percentage terms the other intercrops that grew in the presence of eight of banana's secondary crops. The first column includes intercrop information on all those banana fields that contained chocho. In order to find out whether chocho encouraged or discouraged the presence of other intercrops, column 2, for example, should be compared with column 9. Column 9

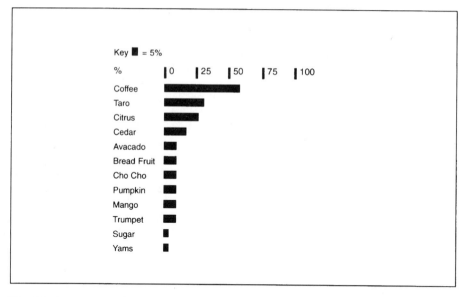

Fig. 30: Intercrops with bananas (limestone area, Christiana, Jamaica)

Secondary Crops	1 Banana–Taro fields (48)* %	2 Banana–Chocho fields (27) %	3 Banana–Pumpkin fields (17) %	4 Banana–Coffee fields (97) %	5 Banana–Avocado fields (25) %	6 Banana–Cedar fields (40) %	7 Banana–Citrus fields (54) %	8 Banana–Pimento fields (25) %	9 Percentage presence of secondary crops in all banana fields (157) %
Taro	100.0	25.9	41.1	27.8	20.0	32.5	35.1	32.0	29.9
Yam	8.3	3.7	11.7	6.1	12.0	2.5	3.7	8.0	8.9
Chocho	14.5	100.0	23.5	14.4	32.0	10.0	24.0	24.0	17.1
Pumpkin	14.5	14.8	100.0	10.3	12.0	15.0	16.6	12.0	10.1
Coffee	58.3	51.8	64.7	100.0	72.0	80.0	75.9	80.0	61.7
Avocado	12.5	29.6	17.6	18.5	100.0	17.5	20.3	32.0	15.9
Cedar	29.1	14.8	35.2	32.9	28.0	100.0	31.4	24.0	25.4
Citrus	41.6	48.1	52.9	42.2	44.0	42.5	100.0	52.0	34.3
Pimento	16.6	22.2	17.6	20.6	32.0	15.0	24.0	100.0	15.9

*number of fields.

Fig. 31: Intercrop interactions within banana fields (limestone area, Christiana, Jamaica)
Note: Fig. 31 gives the percentage occurrence of selected banana intercrops on the presence of other intercrops. The data were collected on small farms in the limestone area west of Christiana, Jamaica. Smaller and short-lived crops are at the top and left of the matrix. Tall, long-lived crops are on the right (not column 9) and at the bottom.

includes banana fields with, and those without, chocho. For example, coffee occurred in 51.8 per cent of all banana–chocho fields (column 2) but in 61.7 per cent of all banana fields (column 9). Chocho vines (which need sun) reduced the presence of coffee (which needs shade) by about 10 per cent. This may mean that chocho and coffee are not compatible crops; it may mean that conditions optimum for chocho vines are not the best conditions for coffee.

For example, the plus and minus numbers in the banana–taro column of Fig. 32 were derived from Fig. 31 by comparing the percentage of the other intercrops in the 48 banana fields which had taro, with the percentage presence of those intercrops in all 157 banana fields. For instance, the figure opposite chocho in the taro column is −2.6; thus, these two crops have an insignificant tendency to discourage the presence of one another. The figure opposite citrus is +7.3; taro and citrus would, therefore, seem to be compatible crops.

In column 2 (banana–chocho), we see (as noted above) that there is 9.9 per cent less coffee in fields with chocho than in all banana fields. Yams (a shorter crop that needs sun) and bananas (a tall crop that needs sun) grew together so seldom that no banana–yam column was constructed.

Fig. 33 is a consolidated version of Fig. 32 which brings out trends more clearly. The table has been shortened vertically by averaging, each together, short crops, vines and trees.

Other Secondary Crops	1 Banana–Taro fields (48)* %	2 Banana–Chocho fields (27) %	3 Banana–Pumpkin fields (17) %	4 Banana–Coffee fields (97) %	5 Banana–Avocado fields (25) %	6 Banana–Cedar fields (40) %	7 Banana–Citrus fields (54) %	8 Banana–Pimento fields (25) %
Taro	–	−4.0	+11.2	−2.1	−9.9	+2.6	+5.2	+2.1
Yam	−0.6	−5.2	+2.8	−2.8	+3.1	−6.4	−5.2	−0.9
Chocho	−2.6	–	+6.4	−2.7	+14.9	−7.1	+6.9	+6.0
Pumpkin	+4.4	+4.7	$\frac{1}{N}$	+0.2	+1.9	+4.9	+6.5	+1.9
Coffee	−3.4	−9.9	+3.0	–	+10.3	+18.3	+14.2	+18.3
Avocado	−3.4	+13.7	+1.7	+2.6	–	+1.6	+4.4	+16.1
Cedar	+3.7	−10.6	+9.8	+7.5	+2.6	–	+6.0	−1.4
Citrus	+7.3	+13.8	+18.6	+7.9	+9.7	+8.2	–	+17.7
Pimento	+0.7	+6.3	+1.7	+4.7	+16.1	+6.3	−0.9	–

*number of fields.

Fig. 32: The net effect of selected banana intercrops on the presence of other intercrops (limestone area, Christiana, Jamaica)
Note: The data were collected from small farms in the limestone area west of Christiana, Jamaica. The percentage figures in columns 1–8 of Fig. 31 were compared with figures in column 9. The resulting plus or minus figures, showing compatibility or non-compatibility, are given in Fig. 32.

	1	2	3	4	5	6	7	8
	Banana–Taro	Banana–Chocho	Banana–Pumpkin	Banana–Coffee	Banana–Avocado	Banana–Cedar	Banana–Citrus	Banana–Pimento
	%	%	%	%	%	%	%	%
Short	−0.6	−4.6	+7.0	−2.5	−3.4	−1.9	–	+0.6
Vines	+0.9	+4.7	+6.4	−1.5	+8.4	−1.1	+6.7	−4.4
Bush	−3.4	−9.9	+3.0	–	+10.3	+18.3	+14.2	+18.3
Trees	+2.1	+5.8	+8.0	+5.7	+9.5	3.0	+5.6	+10.8

Fig. 33: The interaction of selected banana intercrops with crop types (limestone area, Christiana, Jamaica)
Note: The short category of plants includes taro and yams; vines include chocho and pumpkin; bush includes only coffee; and trees, the four remaining crops of Fig. 31: avocado, cedar, citrus and pimento.

Figures for banana–taro fields are very similar to those of average banana fields. Since there are both sun-loving and shade-loving varieties of taro, this short plant can grow under both dense and sparse plantings of bananas and other intercrops.

Chocho (chayote, column 2) is a climbing vine which needs sun. It will tend to climb over and hinder short plants, and its need for sun is the opposite of the shade requirement of coffee bushes. Chocho climbs on trees and does well in fields where sun-loving pumpkin vines are also an intercrop.

Since pumpkins are not such high climbers as chocho, they do better with short crops. They also grow along the edges of wooded fields which explains their presence with bushes and trees.

The presence of coffee does not encourage the presence of short crops or vines, but it does benefit from the shade of trees.

All four tree species either discourage the presence of short crops or have no effect. Coffee is a very important intercrop in banana plots which have trees because it grows below most of the tree leaves. Trees tend to be grown in the presence of other trees since fields devoted to bananas are already committed to tall crops, but there is more competition between one tree and another than there is between trees and coffee.

Shorter sun-loving crops such as maize, kidney beans, Irish potatoes and sweet potatoes occur in only 5.7 per cent, 0.6 per cent and 3.1 per cent of banana fields. (Data omitted from Fig. 31.) Thus, farmers must grow these short crops in other fields with other intercropping associations.

Fig. 34 is a still further compaction of Fig. 32 which shows even more clearly how trees tend to be grown together. Most striking is the positive correlation between trees and coffee bushes. Coffee bushes are encouraged more by the presence of trees (+15.3) than are trees by the presence of other trees (+7.2).

	S	V	B	T
	%	%	%	%
Short	−0.6	+1.2	−2.5	−1.2
Vines	+0.9	+5.6	−1.5	+4.6
Bush	−3.4	−3.5	–	+15.3
Trees	+2.1	+6.9	+5.7	+7.2

Fig. 34: The interaction of crop types arranged by physical size and shape with other crop types in banana fields (limestone area, Christiana, Jamaica)

Bananas (in the conglomerate area)
A parallel analysis of 189 banana fields in the conglomerate rock area east of Christiana has similar results to the previous analysis of banana fields in limestone country. Fig. 35 shows, in column 7, the percentage of banana fields in which each secondary crop occurred. The other six columns show in percentages the effect of each secondary crop on the other secondary crops. Column 3, for example, concerns all the banana fields that contained cacao. Citrus occurs more often in cacao–banana fields than in all-banana fields; thus, these two crops

Secondary crops	1 Banana–Sugar fields (19)*	2 Banana–Taro fields (43)	3 Banana–Cacao fields (27)	4 Banana–Coffee fields (94)	5 Banana–Breadfruit fields (30)	6 Banana–Citrus fields (33)	7 Percentage presence of secondary crops in all banana fields (189)
	%	%	%	%	%	%	%
Sugar	100.0	4.6	–	2.1	3.3	12.1	8.4
Taro	21.0	100.0	11.1	29.7	20.0	33.3	23.2
Yam	5.2	2.3	–	–	3.3	–	3.7
Pumpkin	15.7	4.6	3.7	4.2	–	9.0	4.7
Cacao	–	6.9	100.0	21.2	23.3	21.2	14.2
Coffee	15.7	65.1	74.0	100.0	73.3	63.6	49.7
Avocado	–	4.6	11.1	6.3	13.3	6.0	4.7
Breadfruit	5.2	13.9	25.9	23.4	100.0	33.3	15.8
Citrus	26.3	25.2	25.9	22.3	36.6	100.0	17.4

*number of fields.

Fig. 35: Intercrop interactions within banana fields (conglomerate area, Christiana, Jamaica)
Note: Smaller and short-lived crops are in the top and left hand columns (sugar and taro). Bushes, which are fairly long-lived (cacao and coffee) are in the centre, while trees are on the right (citrus and breadfruit). Yams, pumpkin and avocado occurred in so few banana fields that no columns are devoted to them.

presumably grow compatibly and help to increase total yields instead of lowering them.

In Fig. 36, the two short crops are on the left (columns 1 and 2), two bushes in the middle (columns 3 and 4) and two trees on the right (columns 5 and 6). If all banana fields with sugar are considered, where sugar grows in clumps in open banana fields, cacao which requires shade does not occur, and the presence of coffee is reduced by 34 per cent. The other relatively small crops, taro and yams, are not affected by sugar, while trees, except for citrus, are not compatible with sugar. Citrus is grown both in open fields, which are sometimes used for a few years to grow bananas and perhaps sugar, and in densely-shaded fields with tall crops such as other trees, bananas, coffee and/or cacao. This is why sugar and citrus grow well together, but sugar and other trees such as avocado and breadfruit do not.

Taro is more or less unaffected by the short crops and vines mentioned in the upper part of Fig. 36. The sun-loving variety of taro grows well in open fields with citrus, while shade-loving taro does well in shady banana–coffee fields where there is citrus. Avocado and breadfruit have little effect on the presence of taro but do encourage the presence of cacao.

It thus seems fairly clear, when the cacao column (3) is also examined, that cacao and taro compete against each other. The data indicate that banana–coffee fields have either cacao or taro as a third crop. Cacao reduces the presence of taro by 12 per cent and increases the presence of coffee by 24 per cent. All three trees at the bottom of the table grow well with cacao and provide it with shade. The

	1 Banana–Sugar %	2 Banana–Taro %	3 Banana–Cacao %	4 Banana–Coffee %	5 Banana–Breadfruit %	6 Banana–Citrus %
Sugar	100.0	−3.8	–	−6.3	−5.1	+3.7
Taro	−2.2	100.0	−12.1	+6.5	−3.2	+10.1
Yam	+1.5	−1.4	−3.7	−3.7	−0.4	−3.7
Pumpkin	+11.0	−0.1	−1.0	−0.5	−4.7	+4.3
Cacao	–	−7.3	100.0	+7.0	+9.0	7.0
Coffee	−34.0	+15.1	+24.3	100.0	+23.6	+4.3
Avocado	−4.7	−0.1	+6.3	+1.6	+8.6	+1.3
Breadfruit	−10.6	−1.9	+10.1	+7.6	100.0	+17.5
Citrus	+8.9	+7.8	+8.5	+4.9	+19.2	100.0

Fig. 36: The net effect of selected banana intercrops on the presence of other intercrops (conglomerate area, Christiana, Jamaica)
Note: The percentage figures of columns 1 to 6 in Fig. 35 were compared with the figures in column 7. The resulting plus and minus figures suggest compatibility (in banana fields) of the two intercrops concerned.

reason cacao is not grown in the limestone country west of Christiana is that the crop does poorly in soils derived from limestone.

Banana–coffee fields (column 4) have more taro than the average banana fields and also, as we would expect, more trees and cacao. Cacao and coffee in this area both seem to benefit from shade; one crop does not eliminate the other because replacement intercropping is used so that some potential spots for coffee bushes are occupied by cacao. Pumpkins, sugar and yam are partially or completely discouraged in banana–coffee and banana–cacao fields.

Breadfruit, like cacao, is a much more common crop east of Christiana, as it prefers the somewhat lower elevations (mostly below 2 000 feet) of the conglomerate area. The presence of this crop discourages all the shorter crops and encourages shade-loving bushes and trees.

Since citrus, as explained above, does grow in open fields, this explains the positive correlation with sugar, taro and pumpkin. But citrus also does well with bushes and trees since it is tall enough to get its share of the sunlight.

Fig. 37 is a vertically compressed version of Fig. 36, and Fig. 38 has been compacted in both directions. The data in Fig. 37 show that short plants have no particular effect on other short plants. Sugar, which requires sun, does well with pumpkins, does not co-occur with cacao in banana fields, and greatly reduces the percentage presence of coffee. Sugar seems to tolerate the presence of trees only because the positive correlation with citrus cancels out the negative correlation with avocado and breadfruit (see Fig. 36).

Fig. 38, like Fig. 34, shows that once a field has been dedicated to tall crops like trees and bushes, it is logical to locate more tall crops there than try to include short, sun-loving crops, since tall crops tend to be long-lived and short crops short-lived.

Up to this point, the discussion has centred on whether or not a crop occurred in a field with other crops. The question of how many times each crop is grown in an intercropped field can perhaps be answered best with a map. Fig. 39 shows all the crops in one of Mr V's fields in a limestone valley in 1955. The crops

	1 Banana–Sugar	2 Banana–Taro	3 Banana–Cacao	4 Banana–Coffee	5 Banana–Breadfruit	6 Banana–Citrus
	%	%	%	%	%	%
Short	−0.4	−2.6	−10.0	−1.2	−2.9	+3.4
Vine	+11.0	−0.1	−1.0	−0.5	−4.7	+4.3
Bush	−24.1	+3.9	+24.3	+7.0	+23.3	+5.7
Trees	−2.1	+1.9	8.3	−4.7	+13.9	−9.4

Fig. 37: The interaction of selected banana intercrops with different physical types of crops (conglomerate area, Christiana, Jamaica)

	Short	Bush	Trees
	%	%	%
Short	−1.5	−5.6	+2.0
Vine	+5.5	−0.8	−0.2
Bush	−10.1	+14.5	+11.0
Trees	−0.1	+6.5	+11.7

Fig. 38: The interaction of physical types of intercrops in banana fields (conglomerate area, Christiana, Jamaica)

which this field grew in 1951 are shown on a map published in the *Canadian Geographer*, Vol. 2, 1961, pp. 19–23 (Innis). The central area slopes gently towards a ditch which flows towards the left (west) when it rains. At the top and bottom of the map are steeply sloping sides of a dolina. (A dolina is a completely enclosed valley in limestone country from which rainwater can escape only by going underground.)

While this typical Jamaican small-farm field is the sort that has been described as 'higgledy-piggledy' by writers who have not studied it closely, there is now overwhelming evidence that this is a highly sophisticated space–time crop continuum which tends to make maximum use of available sunlight and precipitation, which provides a constant supply of varied foods, which prevents soil erosion, which discourages weeds from growing, and which accomplishes the multiple goals of good farming practice that intercropping makes possible, as discussed in Chapter 1. This particular field was still devoted to bananas and intercrops more than 30 years after 1951. Other such banana fields in Jamaica are known to have been in continuous production for 50 and even 70 years. The farmers know that fields located in a valley will benefit from hillside erosion. They also keep up fertility by cutting large amounts of brush and leaf mulch from limestone hillsides and hilltops to use as mulch in between the plants. In the 1980s, however, less brush was cut for mulch. By the 1980s, many hill farmers, observed by the author in both Jamaica and Nepal, were using chemical fertilizer. They may find it cheaper to buy mineral fertilizer than to hire people to cut brush, but this practice tends to increase cash outflow from poor countries to purchase fertilizer and makes the agriculture more dependent on non-renewable resources.

The positioning of each plant is done with an eye to plant-neighbours with regard to their time of maturity, the architecture of branch (leaf) and root systems, and state of development. More moveable plants like taro can be replanted each year in different spots, taking into consideration the relatively fixed location of trees and bushes and the less fixed location of bananas and plantain. While the stub of a harvested taro plant can be put back in the same hole the taro came from, or be planted elsewhere, change in location of a banana plant in such a field is usually a choice of suckers on one side or the other of the old plant.

Some calculations of plant spacing in Mr V's field (Fig. 39) have shown that bananas, coffee and taro are all spaced at about the distances recommended by the Jamaican Agricultural Society (JAS). This might show that peasant experiments and agricultural research station experiments arrived independently at similar conclusions. But the small farmer is intercropping, and the JAS is not.

Intercropping in the Christiana area of Jamaica

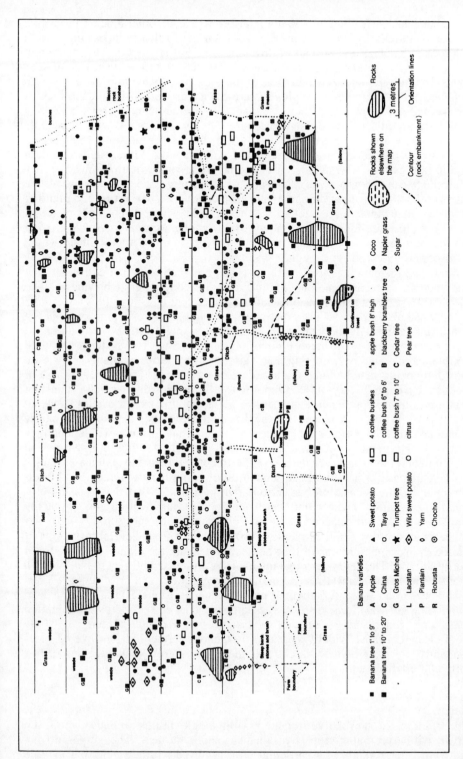

Fig. 39: Map of a field on a Jamaican farm

Since bananas, coffee and taro do not interfere with each other much, they can be spaced as if they were monocrops. The small farmer further reduces competition by growing bananas and taro all of different ages. For an area of 730 square feet (27 feet × 27 feet), the JAS recommends 9 bananas, 9 coffee bushes, or 45 taro plants. In four randomly selected plots of this size, Mr V's average density was $5\frac{1}{2}$ bananas, plus 3 coffee bushes, plus 31 taro plants, as well as 18 banana suckers and 4 small coffee plants. Although banana and taro are annual crops in Mr V's field, they are all of different ages because they are harvested, with immediate replanting, on a daily or weekly basis. Since these crops are of different size, they can be positioned more closely and still give the same yields. Modern agricultural experts who decide on crop spacing assume simultaneous maturity of all the banana or taro crops; therefore, wide spacing is needed because all the plants will be full-sized at the same time. The continuous harvesting and planting which a Jamaican family practises creates a field with plants of all different sizes. The close fitting together of plants of different ages greatly reduces erosion and leaching while continuously preserving a local humid microclimate which helps protect plants from drought.

Analysing the data for three- and four-crop combinations is more difficult than studying two-crop combinations. In limestone plus conglomerate banana fields in Jamaica, 16 per cent had citrus plus coffee, 16 per cent had citrus plus taro, and 5 per cent had all four crops. The creation of compatible intercropping systems is a significant cultural achievement of indigenous small farmers, accomplished during centuries of plant breeding and experimentation. The systems are quite flexible and capable of accepting new varieties and species of plants, which are constantly being searched out and tested. While the small farmers of this area of Jamaica have paid very little attention to the Extension Service's constant pressure to monocrop, when Panama disease-resistant banana varieties (Lacatan and Robusta) were introduced by the Extension Service, the farmers readily abandoned the old Gros Michel variety. Traditional farmers will adopt new practices and planting material if they are helpful, but not otherwise. The rejection of practices advocated by experimental farms and extension agents has given small farmers the false reputation of being narrow-minded and tradition-bound. This accusation has serious ramifications in that it is used by prospective plantation owners and agribusinesses as justification for displacing small farms with aid-financed monocrop plantations of export crops.

The rationality of small farmers is further evident from field observations of intercropping combinations in four homoclimes along the eighteenth parallel of North latitude. The study area in Jamaica is at around 2 000 feet elevation with about 75 inches of rainfall. In a Mexican homoclime near Fortin de las Flores, Vera Cruz, in an Indian homoclime near Panchgani, Maharashtra and in a Thai homoclime in Mai Serai district, the same banana–coffee–citrus combination was found. Since farmers are not advised to practise intercropping, it would seem clear that these widely separated indigenous farmers all discovered independently, by their own research, that these crops grew well together in this particular environment.

Maize

Fig. 40 indicates the relative frequency with which various secondary crops are grown with five major crops by Jamaican small farmers. The graphs do not show how many plants of a particular crop occurred within a field, but they do

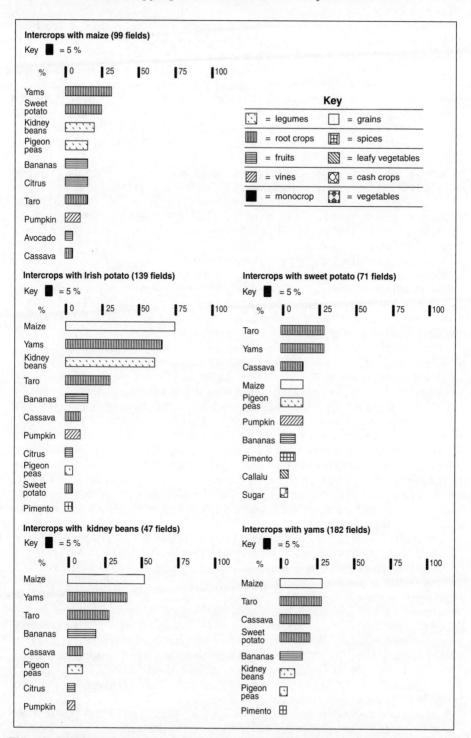

Fig. 40: Intercrops with five major crops of central Jamaica (limestone and conglomerate areas, Christiana, Jamaica)

show how often farmers used a particular crop as a secondary crop in combination with the main crop listed at the head of each graph. Maize (called corn in Jamaica) is not intercropped in 12 per cent of the fields where it is a main crop. Maize grows so quickly and luxuriantly that few smaller intercrops can grow more quickly. Maize is taller than small crops, so when it is closely planted it shades them excessively. Another factor in effect all over the world is that modern hybrid maize is so productive that few other crops can match its yield per acre per month, which means that partial replacement of maize by an intercrop often tends to lower total yields per hectare instead of increasing them.

Yams were grown in 37 per cent of the maize fields, by placing single rows of yam hills about eight metres apart. Maize needs full sunlight, but wide spacing of yam rows and the movement of yam shadows during the day prevents undue shading of maize; if maize and yams are planted at the same time, the slow-growing yams on their poles will not be big enough to cast much shade by the time the maize is harvested.

Since sweet potatoes, kidney beans and taro are too short to shade maize when intercropped, they are accommodated by widening the spacing of maize. Kidney beans utilize environmental resources which young maize is not yet ready to use, and are harvested first. Since sweet potatoes and taro are longer-lived than maize, they begin growing in the maize fields and are then in place to use resources and reduce soil erosion in the period after the maize harvest.

Pigeon peas and cassava are long-lived bushes which eventually grow as tall as maize, or taller, but not until long after the maize has been removed from the field. Bananas, citrus and avocado are taller than maize, but slow growing. Bananas, like sweet potatoes, taro, pigeon peas and cassava can be started in a maize field in order to become a major crop after the maize harvest. Pumpkin and sweet potatoes are both ground-crawling vines whose horizontal leaves are able to utilize sunlight which filters through the vertical maize leaves. These vines never grow taller than the maize. Long-lived fruit trees such as avocado and citrus are widely enough spaced to allow other crops, such as maize, to be grown in the same field. A few fruit trees in a field, which do not cast much shade except in areas right under the tree, will therefore be secondary crops in a whole succession of field plantings with many kinds of major crops and their appropriate associated intercrops.

An analysis of intercrop interaction in maize fields shows very different crop combinations from those found in banana fields.

Fig. 41 analyses only maize fields in limestone country. When columns 1 to 5 are compared to column 6, Fig. 42 is the result. Fig. 42 shows how the presence of kidney beans, for example, affected the other crops. In examining the figures in column 1, it is quite clear that sweet potatoes and kidney beans are incompatible crops. Kidney beans are too short to be able to reach above the creeping sweet potatoes. Usually, kidney beans would be planted along with maize for an early harvest before that of the maize. Sweet potatoes, which tend to be planted later, therefore tend to be in the maize field less often when kidney beans are there. The wide-row planting of yams in maize fields, described earlier, makes it possible to grow kidney beans as well as maize.

The bananas and citrus are tall plants, widely spaced, making it possible for maize to be the major crop for a while. However, the presence of these trees adds enough extra shade to discourage the presence of kidney beans.

In Fig. 42, column 2, it can be seen that the four other intercrops all occur less

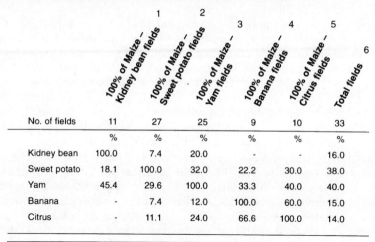

Fig. 41: Intercrop interaction in maize fields (limestone area, Christiana, Jamaica)

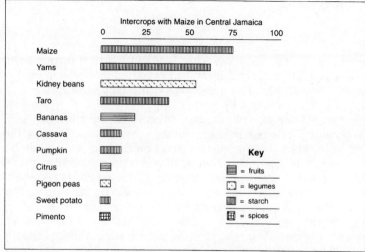

Fig. 42: Intercrop interaction in maize fields (limestone area, Christiana, Jamaica)

frequently when sweet potatoes are present in the maize fields. Sweet potatoes are often started in maturing yam fields; by that time maize has been harvested and is no longer the major crop. Sweet potatoes will probably be important after maize harvest in fields with bananas and citrus, but the presence of these two crops and/or yams, plus maize, provide too much shade and other competition for sweet potatoes.

In yam–maize fields, there is a positive correlation with citrus. Since citrus trees grow for many years, the farmer spaces them widely and does not allow them to preclude the presence of other crops such as yams.

Bananas or citrus in maize fields produce too much shade for the first three relatively short crops in Fig. 42. Since bananas and citrus do not shade each other, these two tall crops tend to be grown together. Since these are, by definition, primarily maize fields, the bananas can be thought of as quite young, or freshly planted suckers. The maize will be harvested before the bananas get tall enough to shade it, while the citrus trees remain as permanent fixtures here and there in the field.

Irish potatoes

Irish potato is shorter than maize but is also a quick-growing, valuable crop. Since maize is a tall, thin crop, it can be a scattered intercrop in fields with many other crops. Fig. 40 has shown that many different secondary crops are intercropped with Irish potatoes in at least 5 per cent of the potato fields. Maize, kidney beans and pumpkin are the only intercrops with a lifespan similar to that of Irish potatoes. The potatoes can be thought of, therefore, as short, quick-growing crops which will produce a good profit from the spaces between the immature, bigger plants. A fair number of maize plants can be scattered here and there in the potato field without lowering yields much, but kidney beans are usually replacement intercrops with Irish potatoes. As they are both bush plants in Jamaica, they don't climb over each other, but as they are also about the same size, one cannot grow under the other. Also, both need full sun. Therefore, a potato plant is omitted from the planting pattern wherever the farmer chooses to grow kidney beans. This replacement intercropping allows a legume to be next to a non-legume.

Kidney beans

Kidney beans are the smallest crop with the shortest time to maturity of the major crops being considered here. This would seem to make it difficult for the farmer to carry on any intercropping in a kidney-bean field, but he or she succeeds by placing rows of maize or yams far apart in spaces where a few kidney beans are omitted. If the tall, thin maize plants are far enough apart, there need be no reduction in bean yield. Since yams and taro take almost a year to mature, they are still quite small by the time of the kidney-bean harvest and have lowered yields only a little. Cassava, pigeon peas, pumpkin and bananas are similarly long-lived plants which can be planted at the same time as kidney beans and which gradually take over the field after the bean harvest. The citrus trees, as we have seen, are a permanent feature, small in numbers, in fields where beans can be grown. Few other trees are found with beans because a field heavily dedicated to tree crops, such as a banana field, is too shady to grow beans.

Sweet potatoes

Sweet potatoes tend to be the follow-up crop in the non-tree fields of the Jamaican hill farmer. A field may be cleared of brush, pasture or weeds and then ploughed up with a heavy hoe. These hill farmers seldom or never use animal or tractor power for ploughing, as their farms are too small and too steep. The fertility of the soil is restored to some extent by grazing a cow or by letting the field go back to brush ('ruinate', as it is called in Jamaica) for a few years.

A field of wild plants has many different species with many kinds of leaf and root configurations. This 'wild intercropping' not only produces a lot of organic matter per year but also reduces nutrient loss so greatly with its extended root systems that the soil quality improves when a field is abandoned to weeds and bushes. This swidden (or milpa, or slash-and-burn) system is widely used all over the world and can make agriculture a non-destructive industry if enough time can be devoted for wild vegetation to rebuild the soil after a human-use period of cropping, harvesting and the higher rates of erosion and leaching which accompany these activities. Intercropping is apparently an attempt to partially reproduce with cultivated plants the soil protection and soil-building which the complicated ecologies of wild-plant associations can accomplish.

When a Jamaican farmer's land is cleared, he or she tends to plant crops that need good soil, such as Irish potatoes or yams. Intercropped Irish potatoes and kidney beans, with wide rows of yams and some taro, may be followed by kidney beans with yams and taro, but no potatoes. By the time of the kidney-bean harvest, the yams and taro plants are quite large. After the beans are removed, sweet-potato slips are planted. The yam harvest, which begins six or eight months after planting, continues for some time, partly because different varieties are grown and partly because yams can be left in the ground to continue growing until the farmers need them unlike Irish potatoes and kidney beans, where the tops all die back and the whole crop must be harvested before it rots. Within six months after their planting, the sweet potatoes, which have been holding the soil in the field as it becomes increasingly open as more and more yams and taro are harvested, will be producing well and will have become the main crop. Small farmers can do things with care that large-scale profit-oriented farmers would never do. For instance, small farmers often feel the dirt around the base of the sweet potato vines to locate tubers which are then carefully broken off, leaving the leaves still connected to the roots. After this operation, the sweet potato plant is still in a productive mode, so that a second and even third crop of sweet potato tubers can be obtained from the same plants. A full system of leaves and roots already in position can produce another crop of sweet potatoes much faster than a new planting of young vines. If a farmer wants to take sweet potatoes to market, he or she will harvest many sweet potatoes on the same day, but if they are for the use of the family, they can be harvested as needed over a long period of time.

Cassava and pigeon peas are relatively long-term bushes which continue growing in fields until sweet potatoes become the main crop. The maize crops indicated in Fig. 42 were probably planted with kidney beans after an Irish-potato plus kidney-bean harvest and remained in fields for sweet potatoes to become the main crop.

Yams

Yams of many varieties are grown in each field. They ripen at different times and have different flavours and different resistance to drought and disease. These large tubers are usually grown in hills and have large, bushy vines which climb on poles. A farmer who needs a lot of yams will make them the main crop. As in Africa, various secondary crops are planted on each hill with the yam vines, while still other secondary crops are grown in the spaces between yam hills. Maize and kidney beans are quick-growing crops which can be harvested from the yam hills before the yam vines are large enough to use most of the available nutrients, light, water and space. Taro, cassava and bananas are year-long crops which can be fitted into yam fields in various ways. Since taro is so short that it will not hinder yams, yam spacing need not be widened to include taro. Cassava and bananas grow quite large, however, and need full sun. Some yams can be omitted from the planting pattern to allow cassava and bananas to be grown. Since sweet potatoes, as we have seen, are intercropped with fully-grown yam vines, they can take over the field as the various yam cultivars have been harvested. Pimento (allspice) is a small tree which grows here and there on the landscape; it is useful in yam fields because poles from several yam hills can be leaned against it.

The almost complete lack of controlled experiments with three or more crops at once make it very difficult to state clearly the benefit which small farmers derive from intercropping. Moreno and Hart's experiment in Costa Rica show that benefits increase as additional crops are put into mixtures, so that the yield of four crops can be almost three times the yield of one crop. The following calculations are based on yields cited by Jamaican small farmers (Fig. 43). A controlled experiment in the Allsides district just east of Christiana, Jamaica, is also cited (Fig. 44). In the following estimation based on orally-reported yields, each plant is assumed to yield the same whether monocropped or intercropped. This method of calculation is known to underestimate the benefit of intercropping, but not enough is known about the subject to be able to evaluate what additional factors should be considered for each plant.

The extra weight of crops in the intercropped field is 3 012 kilograms, which is about 18 per cent extra yield. The widely-spaced yams occupy land which would otherwise be occupied by Irish potatoes and the following crop of kidney

With intercropping (some overlapping of crops)		Without intercropping (each crop follows the previous crop)	
	kg/ha		kg/ha
Irish potato	8 839	Irish potato	9 311
Bean IC with potato	73		
Yam IC with potato	3 413		
Maize IC with yam	29		
Kidney bean	551	Kidney bean	583
Sweet potato	7 051	Sweet potato	7 051
	19 956		16 945

Fig. 43: Estimated yield from one hectare in the course of 18 months (limestone area, Christiana, Jamaica)

	Economic Biomass (tons/ha)	LER relative to MC yam	Net farm income (Jamaican $/ha 1981)	Net fLER relative to MC yam
1. Yellow yam	31.48	1.00	8 920	1.00
2. Yellow yam Bean followed by onion	37.44	1.19	12 055	1.35
3. Yam Sweet maize followed by bean	41.84	1.33	13 477	1.51
4. Yam Grain maize followed by Irish potato	36.74	1.17	9 960	1.12
5. Yam Irish potato followed by radish, followed by African bean	45.29	1.44	14 850	1.66
6. Yam Pumpkin followed by sweet maize	39.84	1.25	12 167	1.36
7. Yam and cabbage followed by carrot, followed by bean	34.25	1.09	9 990	1.12
8. Yam & sweet potato Bean (relay planting)	29.09	0.92	7 578	0.85
9. Yam & cassava and bean	37.69	1.20	12 230	1.37
10. Yam & ginger and sweet potato	41.93	1.33	12 771	1.43

Fig. 44: Results of yellow yam intercropping experiments (Allsides district, east of Christiana, Jamaica)
Source: Schroder and Warnken.

beans. But the reduced yield of these two main crops is more than compensated for by the long-term yam yield. Extra kidney beans with the potatoes, and maize with the yams, give still more yield. In both the intercropped and non-intercropped estimates, the sweet potatoes followed all the other crops and had the whole field to themselves. In a real situation, many of the intercrops would, in fact, be more productive than shown here, because each plant would have more environmental resources available to it since it would not be completely surrounded by other plants of the same species; also, a real field might have one pimento or citrus tree and some taro.

A controlled experiment carried out in the Allsides district of Christiana shows that the measured benefits of intercropping in Jamaica are similar to those of other countries (Schroder and Warnken). In this experiment, yellow yam was intercropped in most cases with only one other crop. Yam-relay cropping with sweet maize followed by kidney beans gave 33 per cent more yield by weight and 51 per cent greater net monetary yield than monocropped yam. Yam, plus ginger, plus sweet potato was almost as productive. Yam plus Irish potato, followed by radish, followed by African bean, gave 44 per cent more by weight and 66 per cent more monetary yield. The experimenters' conclusion is apparently not intended to be humorous when it states that more food and employment would be created if intercropping were 'adopted by even a relatively small percentage of the 220 000 small farms'. This statement is fairly typical of the unwarranted assumptions made by modern scientists about traditional farmers, but it is amazing that the experimenters never noticed the yam intercrop associations within a mile or two of their experimental area. There are areas north of Christiana where yams are monocropped, but this is probably due to the influence of agricultural extension agents.

Secondary crops

When the ten main crops listed in Fig. 29, of which we have discussed six above, are grown as secondary crops, a greater variety of intercropping is possible. When a crop is being treated as a main crop, its peculiarities are respected and secondary crops are chosen which provide little interference with the major crop. But a lot of extra yield of the main crops can be obtained if extra plantings of each crop are made in other fields.

The following tables and discussions do not include those fields where the crop under discussion was a major crop. Fig. 45 shows the number of cultivated species that were observed to occur with various crops. Taro, for example, was observed to be somewhat compatible with 54 other species, but Irish potato and cabbage occurred with only 16 and 17 species, respectively. Taro grows so well with other crops that it occurred as a major crop in only 2 fields and as a secondary crop in 776 fields. The fact that there are both sun-loving and shade-tolerant taro varieties explains why this crop occurs as a compatible intercrop with so many other species.

Yams must have sun but can be widely spaced or grown next to trees which can provide support for the vines. Since most yams are trained on poles, an enormous vine which produces a tuber of 5–10kg or more can be fitted into a small space. Since yam plants are so productive, there is a good incentive to plant them in odd corners or spaces where they will not interfere with other crops. Coffee, which requires shade, occurs with all kinds of tall crops, particularly bananas. Coffee is seldom a main crop, but it is a valuable cash-producing crop which can be grown in any field with tall crops.

Bananas need sun and, as a second crop, can be grown in small numbers in most fields. A banana plant produces a lot of valuable food after a year but is not as appropriate as are yams for fitting into odd corners because the leaves spread out and shade a fairly large area as the plant matures. A banana plant is often grown in an open field with citrus or some other tree, since ground which is already shaded by one tall plant will not become much less productive if there is another tall plant next to it. Maize needs sun, but its relatively tall, thin stature means that it can be a scattered secondary crop in many fields. Chocho is a climbing vine which grows with many kinds of trees but is seldom found in fields of short crops. The last four crops in Fig. 45 are quite short and need sun which makes them relatively unsuitable for intercropping with tall species. The absence of most tall species reduced the total number of species intercropped with kidney beans, sweet potato, cabbage and Irish potato.

Figs. 46 to 53 list most of the important major and secondary crops found in the fields where the ten crops of Fig. 45 were present as secondary crops.

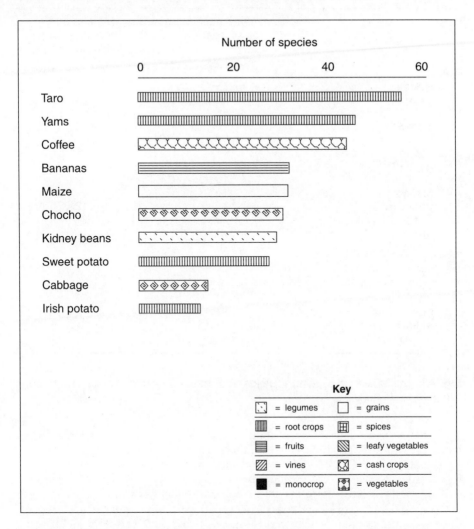

Fig. 45: Number of crop species intercropped with ten common crops where the ten crops were secondary crops (Christiana, Jamaica)

In Fig. 46, some effects of spacing banana plants farther apart can be seen. When bananas are a major crop, they are closely spaced and intercropped with deep shade and shade-tolerant intercrops or crops the same size or taller than bananas (see Fig. 30). When fields with bananas as a major crop are excluded and only fields where bananas are a secondary crop are considered, it can be seen that quite different intercropping associations can be combined with these widely-spaced bananas.

From Fig. 46, it can be seen that the effect of letting in more light increased the number of fields in which taro, sugar and yams were intercrops and decreased the number of fields with cacao, coffee, cedar, mango and trumpet tree. Avocado, chocho, citrus and pumpkin were not much affected. Since taro, sugar and yams need full sunlight, the wider space of bananas as a secondary crop allowed

	Banana as a main crop (close spacing)	Banana as a secondary crop (wider spacing)
	346 fields (%)	260 fields (%)
Coffee	55	41
Taro	28	57
Citrus	25	22
Cedar	16	3
Breadfruit	12	3
Trumpet	11	1
Avocado	10	7
Chocho	10	9
Mango	10	4
Pumpkin	10	12
Cacao	9	1
Sugar	8	18
Pimento	8	13
Yam	7	45
Castor bean	3	8
Sweet potato	2	14
Cassava	2	9
Pigeon pea	2	7
Kidney bean	0.2	14
Irish potato	0.2	11

Fig. 46: The effect of spacing in fields with banana on the choice of intercrops (Christiana, Jamaica)

these crops to be chosen more often as intercrops. An open field with a few bananas is not dedicated to tree crops; thus, cedar, mango and trumpet trees were chosen less often. Coffee and cacao bushes, which need shade, occurred much less frequently, but coffee fits under banana so well that it was still grown in 41 per cent of the fields where bananas were a secondary crop.

Opening fields by growing fewer bananas allows many small sun-loving plants to be grown; these plants were present in less than 5 per cent of the fields where bananas were the main crop. Kidney beans and Irish potato, which grew very seldom in shady banana fields, became significant crops in fields where banana trees were spaced further apart. Sweet potato, cassava, castor bean and pigeon peas also became important crops in such fields.

Chocho, coffee and taro are seldom grown as main crops (Fig. 29). Chocho was a monocrop in 20 per cent of the fields where it was a main crop, but it is a major crop in only 16 fields. In these 16 fields, citrus occurred twice, lima beans once, pumpkin twice and taro once. But as a secondary crop, farmers grew chocho with 31 other species in 76 fields. Trees such as citrus, avocado, pimento and cedar favour the growing of chocho as a secondary crop. Coffee and bananas are other tall crops up which chocho can climb without reducing their yield very much. When chocho is grown intensively as a major crop, many short, sun-loving plants cannot be grown, but when some other crop is the main crop, chocho fits in very well, thus adding an important fresh vegetable to the family's diet at very little extra cost.

Coffee occurred as a major crop in only 14 of the 1000 fields included in this study. Coffee is usually a secondary crop in banana fields; altogether, it was grown as a secondary crop in 306 of the 1000 fields. Fig. 47 shows that bananas, taro, and citrus were the crops most often present in intercropping combinations which included coffee.

Fig. 47 compares the effect of coffee as a main crop and as a secondary crop on the frequency with which farmers chose other crops as intercrops. When coffee is the main crop, farmers must use various trees for shade; when coffee becomes a secondary crop, bananas are the usual main crop. Bananas and taro and often citrus are important combinations where coffee is a secondary crop.

Cabbage is intercropped with more species when it is a secondary crop (see Fig. 48). As explained earlier, cabbage, like Irish potatoes, is a commercial crop in the Christiana area of Jamaica. Since it is intended for sale, big specimens are desirable, and less intercropping is practised. Evidently volunteer taro, which is very common in the country, is often pulled up when the farmer wants to grow cabbage for sale. Pigeon peas and callalu (a variety of amaranth) are not grown where cabbage is a main crop because they produce too much shade.

Maize, like cabbage, is less hospitable to other crops when it is the main crop (see Fig. 49). When maize is densely spaced as a main crop, it is not compatible with the shade from taller crops; thus fewer fields have yams, bananas, and cassava. Trees such as citrus and avocado are relatively permanent fixtures in fields and cannot be removed when the farmer decides to plant maize. Taro probably shades young maize plants too much for it to be grown in maize fields, but where maize is only a secondary crop, taro appears very frequently. Maize is tall enough not to be shaded by sweet potatoes, young pigeon peas and pumpkin; consequently, these are grown more often where maize is the main crop. Because of shade, farmers grow kidney beans less frequently when maize is a main crop, but if kidney beans are the main crop, some maize plants can easily be included.

Unlike maize, Irish potatoes are not a suitable crop to scatter in odd corners of fields. As explained earlier, Irish potatoes are grown primarily to earn cash. This may be the explanation for the similarity between the two columns in Fig. 50. The only crops which seemed to benefit from the wider spacing of Irish potatoes were taro and sweet potato. Few of the other differences are important.

	Coffee as a main crop (close spacing) 14 fields (%)	Coffee as a secondary crop (wider spacing) 306 fields (%)
Citrus	79	28
Banana	57	94
Avocado	57	10
Pimento	43	12
Taro	21	57
Mango	21	10
Breadfruit	14	12
Cedar	14	16

Fig. 47: The effect of spacing in fields with coffee on the choice of intercrops (Christiana, Jamaica)

	Cabbage as a main crop (close spacing)	Cabbage as a secondary crop (wider spacing)
	40 fields (%)	43 fields (%)
Yam	50	65
Maize	38	42
Taro	23	60
Kidney bean	18	26
Cassava	15	23
Pumpkin	13	12
Banana	10	16
Irish potato	8	16
Sweet potato	8	30
Sugar	5	14
Callalu	–	9
Pigeon pea	–	19

Fig. 48: The effect of spacing in fields with cabbage on the choice of intercrops (Christiana, Jamaica)

	Maize as a main crop (close spacing)	Maize as a secondary crop (wider spacing)
	99 fields (%)	418 fields (%)
Yam	37	65
Sweet potato	30	22
Kidney bean	21	32
Pigeon pea	21	11
Banana	14	19
Citrus	13	7
Pumpkin	12	9
Taro	9	63
Avocado	6	3
Cassava	6	15

Fig. 49: The effect of spacing in fields with maize on the choice of intercrops (Christiana, Jamaica)

While kidney beans grow well with Irish potatoes, unlike Irish potatoes they are not primarily a cash crop. Kidney beans were included in fields three times as often as a secondary crop than as a main crop, which is almost the reverse of the ratio for Irish potatoes (see Fig. 51). When kidney beans were grown as a secondary crop, there was a considerable increase in intercropping; this also contrasts with the data for Irish potatoes. Maize, yams and taro, if they are widely spaced, do not interfere too much with kidney beans. However, a tall, shade-producing crop such as bananas is not favoured by wider spacing of kidney beans. The percentage occurrence of the less important intercrops was similar when kidney beans were a secondary crop as opposed to a main crop.

	Irish Potato as a main crop (close spacing)	Irish potato as a secondary crop (wider spacing)
	139 fields (%)	54 fields (%)
Maize	74	69
Yam	63	67
Kidney bean	54	46
Taro	23	82
Banana	19	16
Cassava	12	13
Pumpkin	9	11
Citrus	6	4
Pigeon pea	6	9
Sweet potato	5	13
Pimento	5	4

Fig. 50: The effect of spacing in fields with Irish potato on the choice of intercrops (Christiana, Jamaica)

	Kidney Bean as a main crop (close spacing)	Kidney Bean as a secondary crop (wider spacing)
	47 fields (%)	169 fields (%)
Maize	57	76
Yam	40	69
Taro	28	69
Banana	19	18
Cassava	9	14
Pigeon pea	9	9
Citrus	6	7
Pumpkin	6	10

Fig. 51: The effect of spacing in fields with kidney bean on the choice of intercrops (Christiana, Jamaica)

From the data in Fig. 52 it can be seen that when sweet potatoes are a main crop, they provide a better ecological niche for pumpkin and callalu. Four of the first five crops are much less frequent in fields where sweet potato is the main crop. This is partly a matter of definition because the greater presence of one of these first four crops might make it the main crop instead of sweet potatoes. But 'main crop' is a term used by farmers themselves, and a full crop of sweet potatoes covers the ground so thickly that the potatoes would have a reduced yield if many of the first four crops were planted in the same field. Where sweet potato is a secondary crop, and by definition more widely spaced, then it is possible to add more of other crops as column 2 in Fig. 52 shows. A few relatively long-term crops such as pigeon peas, bananas, pimento and sugar are not affected by the density of sweet-potato planting. Another factor to keep in mind is that sweet potatoes are usually planted as a relay crop in fields

	Sweet Potato as a main crop (close spacing)	Sweet Potato as a secondary crop (wider spacing)
	71 fields (%)	205 fields (%)
Taro	31	52
Yam	30	61
Pigeon pea	17	15
Cassava	15	23
Maize	15	54
Pumpkin	14	5
Banana	10	16
Pimento	8	8
Callalu	7	1
Sugar	7	4

Fig. 52: The effect of spacing in fields with sweet potato on the choice of intercrops (Christiana, Jamaica)

	Yam as a main crop (close spacing)	Yam as a secondary crop (wider spacing)
	182 fields (%)	466 fields (%)
Maize	31	54
Taro	30	65
Sweet potato	19	17
Cassava	18	23
Banana	13	27
Kidney bean	9	24
Pigeon pea	7	9
Pimento	7	9

Fig. 53: The effect of spacing in fields with yam on the choice of intercrops (Christiana, Jamaica)

where initially some other crop is the main crop. As the main crops are removed from the respective fields, there is a transfer of some fields from column 2 to column 1.

Yams as a major crop also have the effect of reducing the number of fields in which maize and taro are chosen as intercrops. Widely-spaced yams, in rows eight yards apart or as scattered single hills, leave a lot of environmental resources (light, water, nutrients, space) available for other crops, so farmers grow markedly higher percentages of maize, taro, bananas and kidney beans in these fields. Long-term crops which are about the same height as yams, and therefore avoid too much shading one way or the other, are cassava, pigeon peas and pimento. Their percentage presence in fields with yams is not greatly affected by the density of yam planting.

Soil fertility

In order to test the effect of Jamaican indigenous intercropping on soil fertility, samples were taken from 13 Jamaican fields in 1961 and again in 1971. Figs 54, 55, and 56 show the use made of fields by three different farmers over a period of 20 years. Generally speaking, there was no marked change in fertility over a ten-year period.

Maintaining the level of organic matter in the soil is done with mulch, brush from hillsides, kitchen wastes, night soil and fallow with brush or grass, or by grazing a cow on grass. Organic material is high (above 5 per cent) in all but two fields. The trend in organic matter over the ten-year period was down in five fields and up in four, two of which were banana gardens right beside the house. High humus levels are important in the tropics to prevent soil laterization; low levels may allow nodules or layers of insoluble iron to form, which may permanently damage the soil. As far as could be ascertained, mineral fertilizer was not added to these fields before 1971, except in a couple of cases. On Mr. R's farm (Fig. 56), field P98 had been treated with purchased fertilizer, which markedly improved the phosphorus content but greatly reduced the level of organic matter.

The nitrogen level is high in the ten fields that do not belong to Mr R. Two of his fields experienced more than a 0.05 per cent N decrease in ten years because he was cropping them heavily and not doing enough to restore fertility. Mr R is wealthier and more cash-oriented than the other two farmers and is more likely to follow the recommendations of agricultural extension agents, but he has not given up intercropping.

Phosphate levels are low to medium-low in most fields. Mr O's P84 field (Fig. 54) has consistently high phosphorus levels because it is beside his house and receives organic wastes. Field P85 is farther from the house and was sold to a mechanic in 1967. The mechanic grew only bananas, without mulch, intercropping or mineral fertilizers, which reduced the phosphorus level to zero by 1971. Mr R's field, M24, had phosphorus fertilizer applied before 1961, which explains why the level of P then was so high and dropped to a more usual level ten years later. Phosphates are a severe problem for farmers in this area, but in most of the fields discussed here, the problem was being handled adequately.

Potash content in the soil is high because it is a common element in the limestone bedrock. Of the 26 soil analyses, only three had below-average K (140ppm). These three were on the land farmed intensively by Mr R.

In effect, mineral fertilization with N, P and K is combined with a dependence on organic matter for all the trace elements plants need. Thus, the addition of the three minerals, which previously had been limiting factors, could not maintain organic levels in the soil indefinitely. Dividing the land into small fields which would be too small for animal- or tractor-powered ploughs but which could be ploughed with hoes, and rotating crops in all these fields greatly helps to retain soil and reduce leaching.

Each of Figs 54, 55 and 56, concerns a different farm, so the timetable by which each farmer brought fields into cultivation and let other fields revert to pasture or brush can be seen. Each farmer has a 'permanent' banana garden to supply the fruits, vegetables, coffee, firewood and lumber which are produced in such fields; each also has fields for crop rotation to provide his family with food from shorter, sun-loving crops. Intercropping in these fields and gardens helps prevent loss of soil and nutrients. The crops mentioned in Figs 54–6 are not the

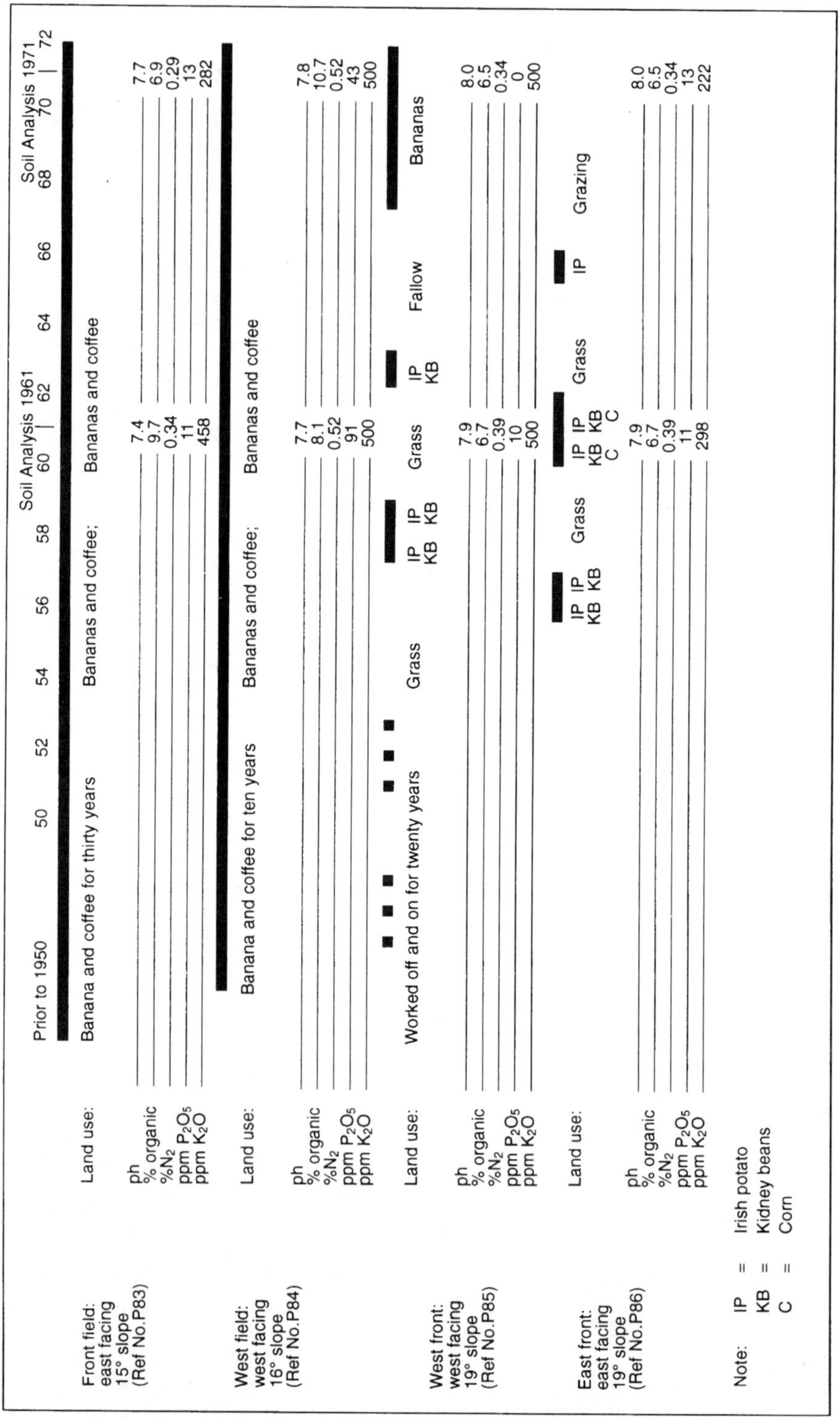

Intercropping

	Prior to 1950	50	52	54	56	58	Soil Analysis 1961 60	62	64	66	68	Soil Analysis 1971 70	72
Front field: east facing 15° slope (Ref No.P83) Land use: ph % organic %N$_2$ ppm P$_2$O$_5$ ppm K$_2$O	Banana and coffee for thirty years			Bananas and coffee				Bananas and coffee 7.4 9.7 0.34 11 458				Bananas and coffee 7.7 6.9 0.29 13 282	
West field: west facing 16° slope (Ref No.P84) Land use: ph % organic %N$_2$ ppm P$_2$O$_5$ ppm K$_2$O	Banana and coffee for ten years			Bananas and coffee;				Bananas and coffee 7.7 8.1 0.52 91 500			Bananas	7.8 10.7 0.52 43 500	
West front: west facing 19° slope (Ref No.P85) Land use: ph % organic %N$_2$ ppm P$_2$O$_5$ ppm K$_2$O	Worked off and on for twenty years			Grass	IP IP KB KB	Grass	Grass 7.9 6.7 0.39 10 500	IP KB	Fallow			8.0 6.5 0.34 0 500	
East front: east facing 19° slope (Ref No.P86) Land use: ph % organic %N$_2$ ppm P$_2$O$_5$ ppm K$_2$O					IP IP KB KB	Grass	IP IP KB KB C C 7.9 6.7 0.39 11 298	Grass		IP	Grazing	8.0 6.5 0.34 13 222	

Note: IP = Irish potato
       KB = Kidney beans
       C = Corn

Fig. 54: Mr O's farm – field histories and soil fertility (Christiana, Jamaica)

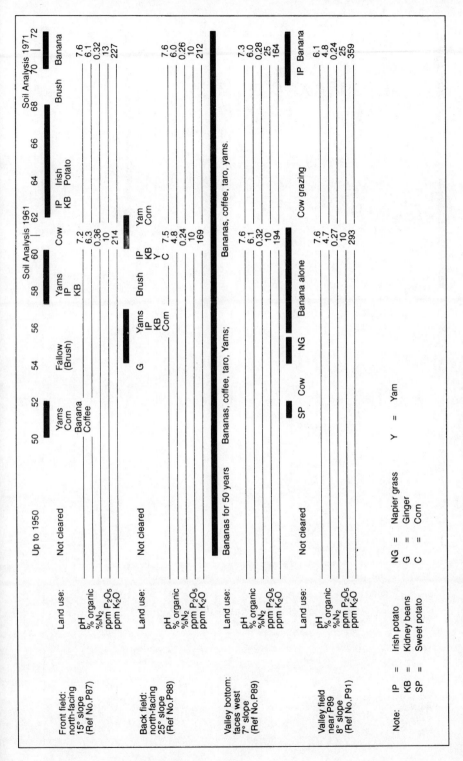

Fig. 55: Mr V's farm – field histories and soil fertility (Christiana, Jamaica)

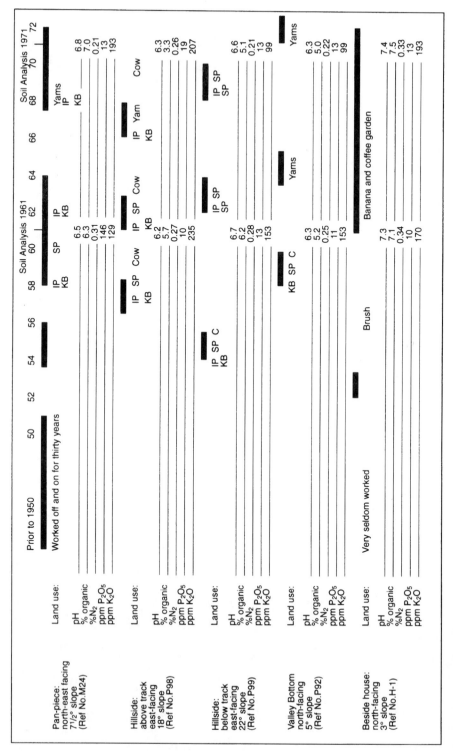

Fig. 56: Mr R's farm – field histories and soil fertility (Christiana, Jamaica)

only ones grown in each field, but they are intended to refer the reader to the earlier discussion in this chapter of the crops grown with each main crop.

The banana–coffee–taro fields are impressive. Mr V's field, P89, mapped in Fig. 39, contained fertile usable soil after 70 years of continuous cultivation. Ten years later, in 1981, the field was still devoted to bananas and coffee. Field P91 was rather neglected in 1981, with less intercropping as such fields can become root-bound in time with little room for taro. In 1981, field P91 was receiving less mulch and inadequate (according to agricultural extension standards) applications of minerals N, P and K. But since Mr V's wife was beginning to cultivate parts of the fields carefully and intensively again like other fields, P91 may go on being productive for another 80 years. Banana fields like this, and gardens close to the house (which have no main crops), are kept productive with high soil quality by small farmers all over the tropical world. Since most of the crops grown in such fields are eaten by the farm family, most of the organic matter never leaves the farm. Thus, in small farm societies, there is no problem of high fuel costs to transport sewage from urban areas back to the land.

From this Jamaican case study it can be understood how small farmers, by means of their indigenous techniques using intercropping, are able to maintain soil fertility in both permanently cultivated plots and also in hillside fields where land use alternates between cultivation and pasture or a period of wild vegetation. In many Third World countries, small farmers have now been excluded from rich, flatter lands by commercial plantations of export crops. Many small farmers who have been driven on to hillsides where they have no terraces, because they have not had the time or energy to build any, and where there are inadequate amounts of land to allow the soil to rest and recuperate, are allowing deforestation and soil erosion to take place. These refugee farmers are not in a position to take much advantage of the agricultural accomplishments of traditional farmers, and their destructive ways should not be considered typical of indigenous methods.

3 Intercropping in Nepal

INVESTIGATIONS CARRIED OUT in the Kathmandu region of Nepal in the autumn of 1980 revealed extensive use of intercropping. This chapter describes the crop combinations used and discusses the effects of terrain, field size and crop type.

Influence of terrain

Immediately north of Kathmandu, there was more intercropping on the sloping fields. On sloping terrain, the average number of crops per field was 1.8; on flat land the average number was 1.2. This is because more plants are necessary to hold the soil of sloping fields and perhaps because poorer farmers who have less land and more need to intercrop have been relegated to sloping land. In the autumn of 1980, when these observations were made, there was less contrast between these numbers than in the early part of the rainy season. In spring and early summer, the hillside maize fields were intercropped with amaranth and several kinds of legumes, and the flat lands were devoted to monocropped rice.

As shown in Fig. 57, Irish potato was the most widely-used autumn crop in the flat fields, and it was the most extensively intercropped. Big Chinese radishes alternated with potatoes in the rows in more than 30 per cent of the potato fields, and turnips were almost as prevalent. The leaf systems of these two intercrops were comparable in size but more vertical than the potato-leaf system. They did not shade the potato nor were they shaded by it.

Garlic is so short in stature that it is hard to find an intercrop that will not shade it too much. In India, garlic was often monocropped; in the area north of Kathmandu it was intercropped with small coriander and spinach plants which cast little shade. Radish was grown less often with garlic than with potatoes.

Rayo, or broad-leafed mustard, a type of spinach, was grown mostly on flat land. It was intercropped in half the fields where it grew, on both flat and sloping land. Large radishes as a main crop were intercropped twice as often on sloping as on flat land.

The area just north of Kathmandu was used for market gardens. The relative lack of intercropping in these flat fields may have been due partly to the use of monocropping to produce larger vegetables for market. Where broad bean, cauliflower, onion and radish were the main crops in these wide, level terraces, more than 50 per cent of the fields were monocropped.

The terraces had held standing water for the rice crop earlier in the growing season, but after the rice harvest, as the monsoons tapered off, the fields became dry enough and well-enough drained for vegetables. The soil seemed excellent and the crops healthy and productive.

On the sloping fields, relay planting insured continuous protection by leafy umbrellas to shelter the soil from raindrop impact and by root structures to hold the soil and reduce erosion. Fig. 58 shows intercrops in sloping fields north of Kathmandu. Before the maize harvest, finger-millet seedlings were transplanted into spaces between the maize plants. After the maize and intercrops were harvested, the thickly planted finger millet remained as a monocrop until its maturity. Only the seed heads of the millet were cut off, leaving the green stems

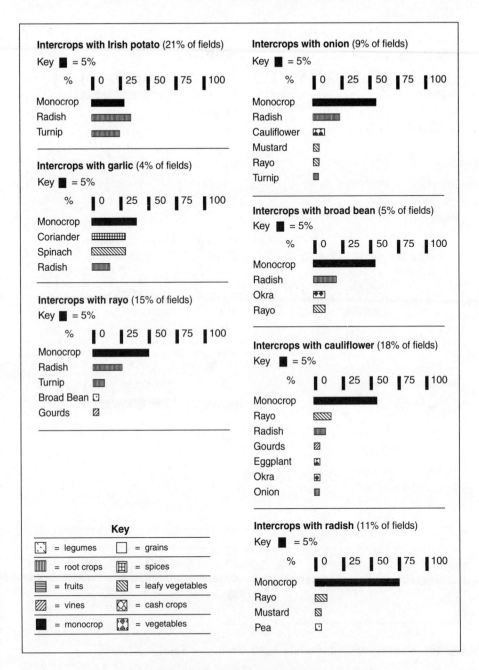

Fig. 57: Intercrops in 406 flat fields (north of Kathmandu, Nepal)

and leaves for animal grazing. It can be seen from this brief description that in Nepal, as in Jamaica, at the cost of more work and less immediate profit, agricultural practices are carried out to insure the maintenance of soil quality into the indefinite future.

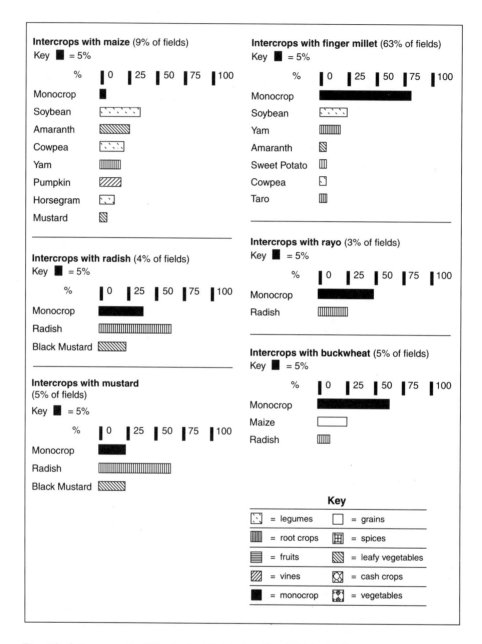

Fig. 58: Intercrops in 403 sloping fields (north of Kathmandu, Nepal)

Although maize fields had been partially harvested by the autumn of 1980, it was observable that maize was almost always intercropped. Examples of legume intercrops were soybeans, cowpeas and horsegram, each of which occurred in more than 10 per cent of the maize fields. Yam and pumpkin were equally common; both these crops were ground vines which did not shade maize but were able to capture and utilize the sunlight not used by the vertical maize leaves. The legumes, also, were somewhat shorter than the maize plants. Red amaranth was a spindly plant almost as tall as maize but planted further apart to reduce competition. For farmers who have not much land and who put a lot of work into hand digging the fields, it is logical to fit as many plants as possible into a field which has a tall, non-bushy crop, such as maize, for the main crop. One former maize field had the identifiable remains of six different legume species.

This glimpse of the work of Nepalese farmers suggests a great deal of sophistication. They are able to feed their families while maintaining the soil in order to feed the family in the future. The beautifully grassed waterways and the long gently-sloping ramps which lower rainwater run-off non-destructively from hillsides down to stream level are further evidence of their concern for the future.

Influence of field size

Kathmandu Valley, which is at 4 000 feet elevation, has a temperate climate, with frost in winter and a winter dry season. Because of the climate, year-long crops apart from trees cannot be used in intercropping. In spite of this, Nepalese farmers show that temperate-land intercropping is quite possible. Their techniques still preserve many of the ancient agricultural traditions which have maintained the soil and fed the people for so long. The range of mountains between Kathmandu and India makes the importation of tractors and gasoline very expensive. Tractors were tried in the valley for about five years and then abandoned, with the exception of some Japanese two-wheeled tractors which are still in use. Some mineral fertilizer is imported and used. Even small-scale farmers sometimes buy mineral fertilizer with their hard-earned cash; they have evidently discovered that money spent on N, P and K will produce more food than the same money spent directly on food in the market.

Roads being built in Nepal tend to change the old kinds of agriculture and social co-operation. When a road comes in, big farmers tend to evict tenants and foreclose on loans in order to have more land for growing commercial crops with seasonal labour. It seemed likely that less intercropping would be practised on the big fields thus farmed. To test this hypothesis, a study was made of large and small fields south of Bhaktapur, in the Kathmandu Valley, about fourteen miles east of the city of Kathmandu via the Chinese electric trolley. A small field was defined as being less than 25 metres square. The 249 large fields averaged 1.92 crops per field, the small fields 2.27 crops. No proof was collected to show that big fields belonged to big farmers and small fields to small farmers. It may well be that all farmers intercropped vegetables on small plots and monocropped rice in big fields. However, small fields formerly in rice were more often cultivated with intercropping than were big fields formerly in rice.

Most of the rice fields included in this survey actually had no crops since the study was done in the autumn after most of the rice had been harvested. Fig. 59, shows that far more large fields than small ones had rice. The figure also shows important differences in the main crops chosen for large and small fields.

Once the main crop was chosen, it was more extensively intercropped in the small fields (see Figs 60 and 61). Chilli, for example, was monocropped in 30 per cent of the large fields where it was the main crop but was never grown alone in small fields. Ginger, a commercial crop, was the leading intercrop in

	Large fields	Small fields
	249 fields (%)	346 fields (%)
Rice	51	18
Ginger	16	4
Mustard	14	20
Chilli	8	9
Cauliflower	3	5
Maize	2	2
Radish	2	4
Black mustard	1	5
Garlic	1	5
Rayo	–	17
Onion	–	3
Pumpkin	–	2
Turmeric	–	2
Broad bean	–	1
Pea	–	1
Finger millet	–	1

Fig. 59: Main crops in large and small fields (% of fields in which each crop was the main crop) (Bhaktapur, Nepal)

large chilli fields, but maize, a food crop, was the leading intercrop in small chilli fields.

Ginger was a more important main crop in large fields than in small fields and was presumably being grown commercially since there is a limit to the amount of ginger a family can consume. Too few small ginger fields were observed to give meaningful percentages of secondary crops, but cowpeas, cluster beans, eggplant, maize, mustard, pumpkin and trees each occurred once in small ginger fields. Small farmers, also, were probably growing ginger as a commercial crop.

The amount and kind of intercropping with mustard was about the same in large and small fields. Mustard and black mustard, sown in the autumn fields after the harvest of some previous crop, were usually dried and consumed during the winter. Great loops of drying braided mustard greens could be seen hanging from south-facing windows of houses all over the valley at this time of year.

Rayo (broad-leafed mustard) is a short crop with spreading leaves which does not fit well into intercropping patterns. Both here and north of Kathmandu, it was frequently monocropped. Garlic was more intercropped here than north of the city. Cauliflower was grown on only 3 per cent of large fields but was somewhat more intercropped here than on the small fields.

The general conclusions from the Bhaktapur study are that there tended to be more cultivation in the late fall on small fields than on big fields; there was a greater variety of main crops on small fields; and usually more intercropping.

Influence of crop type

South of Kathmandu, between the city and the village of Kirtipur, the intercropping patterns in 2108 fields were recorded. This was the first area studied in the autumn of 1980, and here no distinction was made between flat and sloping or large and small fields. This study was designed to find commonly used inter-

Intercropping in Nepal

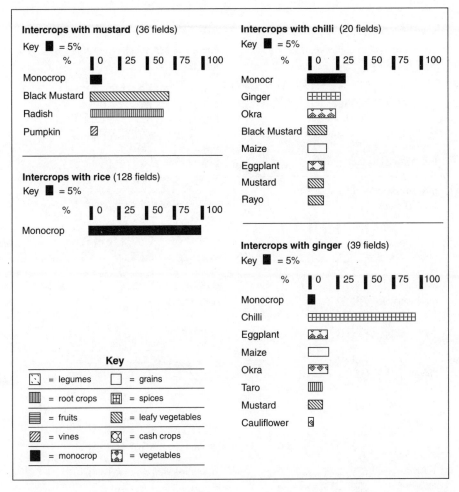

Fig. 60: Intercrops in 249 large fields (more than 25m^2) (south of Bhaktapur, Nepal)

cropping combinations rather than to estimate the frequency with which main crops were chosen and the acreage devoted to them. The sample chosen consisted of all fields in the area with the exception of rice fields. Hundreds of terraced harvested rice fields were omitted from the fields tabulated, but all other kinds of fields were included. It should also be kept in mind that earlier in the year many of these fields were quite different with much more maize intercropped on the gently sloping hillside fields.

Of the grains, rice was the most common and the least intercropped. Because the rice was grown in standing water in order to control weeds, intercropping was done only on the field borders, where the Nepalese sometimes grew a legume called blackgram (see Fig. 62).

The remnants of 144 maize fields, which had not been replanted with a second crop, were studied for evidence of intercropping. All but four of the fields had

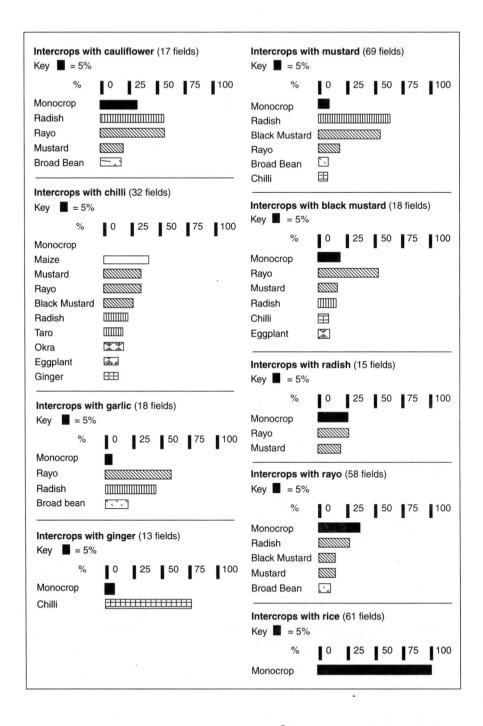

Fig. 61: Intercrops in small fields (less than 25m^2) (south of Bhaktapur, Nepal)

Intercropping in Nepal

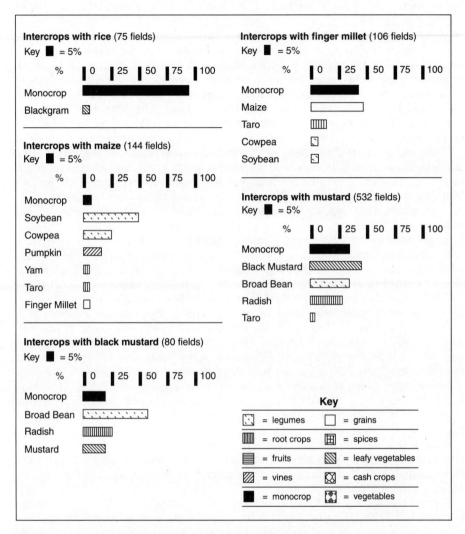

Fig. 62: Intercrops with grains and mustards (south of Kathmandu, Nepal)

been intercropped; soybeans were present in half, and cowpeas were quite common.

Finger, or African, millet, is a much shorter crop than maize. Because it was thickly planted among maturing maize plants, there was little space for other intercrops. It is probable that none of the finger millet fields was monocropped originally. Cowpea and soybean in these millet fields were left over from the maize period. Taro, however, was actively growing in the former maize fields. Since taro, a volunteer crop in both Nepal and Jamaica, has useful edible roots, it was not destroyed when it came up. These fields were so meticulously weeded, in order partly to feed the pulled-up weeds to goats and chickens at the homestead, that decisions had clearly been made to leave alone the four or five taro plants found growing in some of the millet fields.

Many of the autumn fields south of Kathmandu were yellow with flowering mustard. More than one quarter of fields had mustard as the main crop, grown not for seed but for greens. A variety with dark leaves, called black mustard, grown in many fields as an intercrop, provided genetic variation and helped to control damage by insects and diseases. There were also examples of mustard, radish and broad bean intercropping, all in the same field. The mustard plants, which had a tall, spindly form, neither shaded, nor were shaded by, the other two crops.

Since the area studied south of Kathmandu is as close to the city as the northern study area, it is also used for market gardens to supply the city. South of Kathmandu, as in the area north of the city, there was less intercropping with Irish potato, but a great deal of radish was intercropped in potato fields. The turnips, which were popular in potato fields north of the city were, however, not popular here.

Small farmers are very aware of market prices and of the availability of new crop species or varieties and are always trying new crop combinations in keeping with their perceptions of the opportunities which are available. Radish, for example (see Fig. 63), which was monocropped in 75 per cent of the flat fields and 35 per cent of the sloping fields north of Kathmandu, was monocropped in only 40 per cent of the fields south of the city. The southern sample includes both sloping and flat fields. The figure for mustard as an intercrop south of the city was also in between the figures for flat and sloping fields north of the city. Broadbeans were the most popular radish intercrop south of Kathmandu but did not reach the 3 per cent level north of the city.

Broad beans were planted in holes made by a digging stick in fields which had held rice during the summer. Broad beans, which were seldom grown on sloping fields, had radish as the most important intercrop on both sides of the city.

Cauliflower was intercropped in twice as many fields on the south side of the city as on the north side. Both north and south of the city, radishes and rayo were the principal intercrops, but the four next intercrops found with cauliflower were entirely different north and south.

Pumpkin and choyote were grown on small pieces of land, such as angles between big fields. Because most of the choyote grew on trees, taking up little ground space compared to pumpkin, there was small opportunity for choyote intercrops other than trees.

Spices and spicy foods can be grown as intercrops by a small farmer but were also grown as main crops, even though only a little land could be devoted to them (see Fig. 64). Turmeric was monocropped in most fields where it was the main crop. It was a minor secondary crop in chilli and pumpkin fields south of the city.

Onion and garlic are relatively short plants which need full sunlight if they are to produce useable bulbs. Intercropping with them was therefore limited; onions were usually monocropped. Anise, a small, feathery-leafed plant, was a crop observed only in onion fields. Coriander, a small branchy plant without big leaves, was a useful intercrop with garlic and was found in some other fields north of the city.

Chilli, a larger spice plant than garlic and onion, was grown with many intercrops. Chilli has much smaller leaves than turmeric and, unlike turmeric,

Intercropping in Nepal

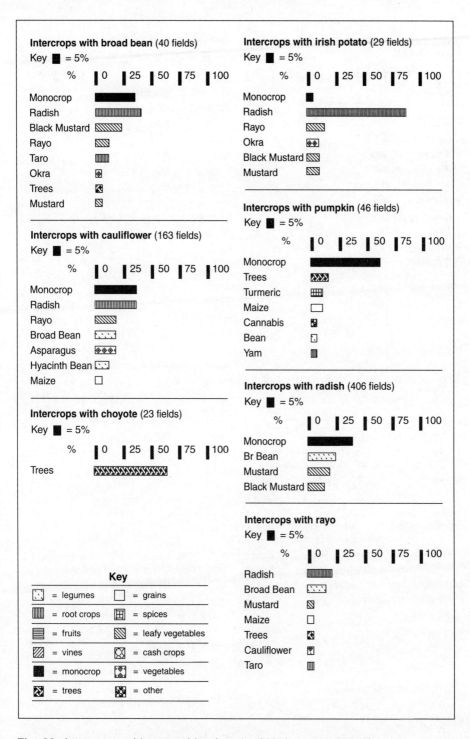

Fig. 63: Intercrops with vegetables (south of Kathmandu, Nepal)

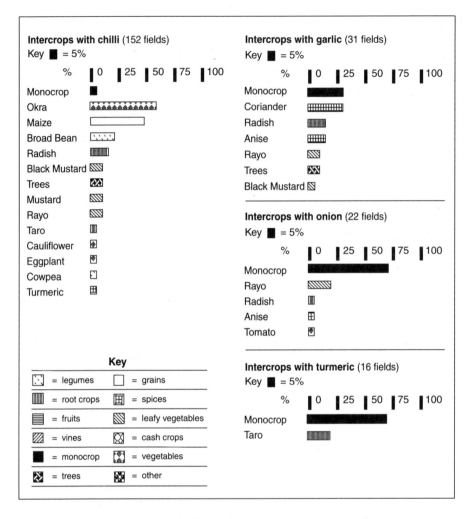

Fig. 64: Intercrops with spicy foods (south of Kathmandu, Nepal)

is harvested over a long period of time. For these reasons, there was much more intercropping with chilli. Okra and maize, taller crops than chilli but not spreading plants, were scattered here and there in the chilli fields and did not much hinder the main crop.

4 Intercropping in India

INDIA'S RICH TRADITION of intercropping is at least as developed as that of Jamaica or Nepal, though many of the crops are different. This chapter reports on intercropping field work in three locations in India, one in the state of Kashmir and two in the state of Maharashtra. Of the latter two, the Alandi study took only a few days, whereas the Girvi investigation took many weeks in each of four different years. Planting patterns will be illustrated by maps, and crop combination adjustments for different elevations will be discussed.

Srinagar, Kashmir

Zero Bridge crosses the River Jhelum immediately south of Srinagar, Kashmir. The market-gardening area studied is south of the bridge, east of the Zero Bridge Road, and within the meander of the river. The field sample included all the fields between the river and the road. Many of the small farmers had other jobs as clerks and typists. Many spoke English and explained why they found intercropping worthwhile.

Collards, radish and kohlrabi were usually monocropped when they were main crops, but there were a few intercrops, as Fig. 65 shows.

Between 25 per cent and 50 per cent of the chilli-pepper fields were monocropped. Where chillies were the main crop, orchard trees, such as apples and pears, were popular intercrops. But where trees were the main crop, they produced so much shade that chillies could not be grown. Fig. 65 shows the importance of eggplant and trees in chilli fields, with some tomatoes and radish also being chosen as secondary crops. The shady fields where trees were the main crop had radish and cauliflower as the most popular intercrops; about 15 per cent of these fields had eggplant or kohlrabi.

The main crop that was most often intercropped was asparagus. The local perception seemed to be that since the asparagus harvest was early in the spring, the field could be used the rest of the year to grow other needed crops. The feathery leaves of asparagus do not seriously shade other crops, and if the intercrops are not too thickly planted apparently they do not reduce asparagus yields too much. The modern versus traditional attitudes torwards gardening seem quite clear here. The modern farmer wants the biggest possible asparagus, either big specimens and/or the biggest yield of asparagus per acre. The traditional farmer wants the biggest possible total yield from all crops. Several Kashmir asparagus fields had both maize and tomatoes as intercrops. Radish or cauliflower were grown in more than 25 per cent of all asparagus fields, and a few fields had all four intercrops with the asparagus.

Alandi, Maharashtra

Another sample of traditional intercropping combinations was taken south of the village of Alandi, which is 12 miles north-east of Poona in the state of Maharashtra. This was not an area producing market vegetables for the city, although there were a few fields of monocropped flowers being grown commercially. The main purpose of the Alandi study was to compare the field results with the

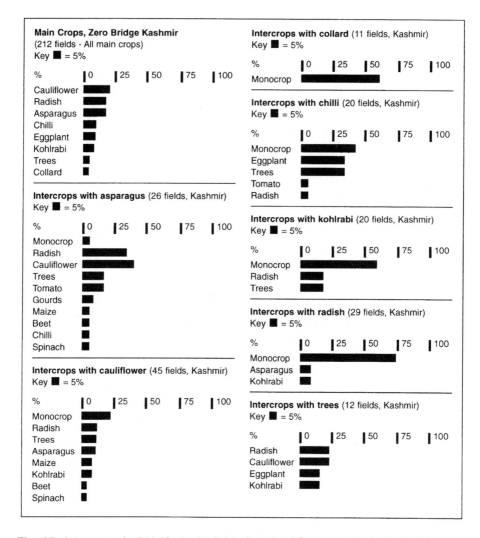

Fig. 65: Intercrops in 212 Kashmiri fields (south of Srinagar, Kashmir, India)

author's principal field study area near the village of Girvi, 60 miles south of Poona. The intercropping patterns chosen by the farmers of the two areas were similar.

Bajri, or pearl millet (see Appendix 2 for Latin names), also known as birdseed, is an important food crop in dry parts of India. Some of the tabulated bajri fields in Fig. 66 were irrigated and some were not. In both Maharashtrian study villages, the intercrops chosen for bajri were different when the bajri was irrigated than when it was not irrigated. In the Alandi area most of the irrigated bajri was monocropped, while most of the non-irrigated bajri was well intercropped. Blackgram, a small legume seldom grown in the Girvi area, was an important Alandi intercrop. The other bajri intercrops shown in Fig. 66 were all legumes and all important in the Girvi area as well.

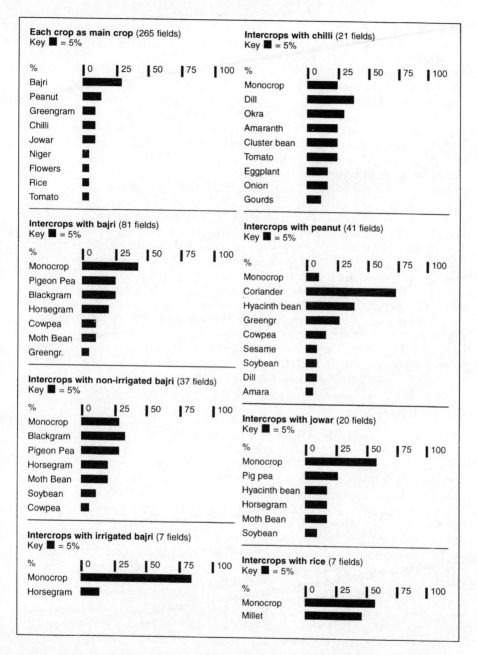

Fig. 66: Intercrops in 265 Alandi fields (south of Alandi, Maharashtra, India)

There was less intercropping in the Alandi area than near Girvi, perhaps because more of the land was owned by big landlords. Land ownership is hard to determine in India because it is often illegal to own a large amount of land, but field size can be a guide to the relative importance of big landowners. Field size was not measured in Alandi, but there seemed to be a higher proportion of big fields than around Girvi.

Rice in the area was quite frequently intercropped with finger millet, which is an unusual combination. If local people have developed a method of growing two crops together where one grew before, it might have important implications for agriculture in other parts of the world. The other important grain of the Alandi area was an Indian variety of sorghum named jowar. Most of the jowar fields were monocropped, which differed from those of the Girvi area, but five legume intercrops were important. The fields were either monocropped, or they had three or four intercrops; it was unusual for these fields to have only one or two intercrops. This would seem to indicate that a farmer decides either to monocrop or to intercrop. Such a decision might well be related to the amount of land owned. Farmers who have only a little land can maximise their yields by intercropping.

Peanuts and coriander seemed to be a workable combination. Coriander is such a short plant that it produces very little shade, and peanuts do poorly if a field is shady. Peanuts are legumes, but as they need sun, they were not grown as secondary crops in jowar or bajri fields. Because of their sensitivity to competition, the peanut needs to be grown as a main crop and intercrops chosen with peanut needs in mind. Soybean, cowpea and hyacinth bean are plants which are taller than peanuts and have bigger leaves. As they are not big enough to cast long shadows, they can be grown in peanut fields if they are widely spaced. Pigeon peas cast too much shade to be grown with peanuts. Dill and sesame are quite spindly plants which cast little shade and which a farmer does not need in large amounts for family use. The small farmer will choose to plant them in a field where a few spice or oil plants will do little harm.

Girvi, Maharashtra

The village of Girvi is about 60 miles south of Poona in the state of Maharashtra. A lava bedrock forms a flat plain north of the village, but two or three miles to the south there are flat-topped lava plateaux. The Girvi elevation is about 2000 feet above sea level; the first flat plateau south of the village (Palwan 1) is at 3500 feet; the second and higher plateau (Palwan 2) is at 4000 feet.

The average rainfall near the village is about 14 inches, but year-to-year variations are considerable. Rains were normal when observations were made in 1966 and 1968 (see Fig. 67), but in 1970 the rainfall was so limited that few of the customary bajri fields were present at the Girvi elevation. In 1980, the rains were normal, and bajri was again grown on the non-irrigated land south of the village.

The village of Girvi is at about the point in the plain where the main stream of water from the hills sinks into the ground. Since the lava bedrock is quite close to the surface in the whole area, the water-table has little water storage capacity. During heavy rains, the streams become rivers, and there is plenty of water to fill the shallow water-table. However, during most of the year, there is no surface water north of the village, nor is there much stored ground water. Southeast of Girvi, a dam has been built which stores enough water for irrigation purposes for part of the dry season. Since the dammed lake often runs out of water a month before the monsoon begins, the supply of irrigation water is not reliable. For wealthier farmers, the water supply may be quite reliable, but poorer people have a less reliable supply, and most of the cultivated land is not irrigated. The two

	1970 (dry year)	1980 (normal rainfall)
	1 222 fields (%)	877 fields (%)
Bajri/millet	7	57
Banana	2	2
Chickpea	2	0
Chilli	5	4
Cotton	4	14
Jowar/sorghum	46	1
Maize	0	1
Onion	2	1
Peanut	3	1
Rice	0	1
Sugar	15	8
Wheat	7	2
Other	7	8

Fig. 67: Frequency with which Girvi farmers chose main crops (Girvi, Maharashtra, India)

plateaux south of the village have no irrigation water, but they do have a little more rain.

Intercropping patterns
Since banana plants live for more than a year and cannot withstand drought, there are few banana fields in this area. Where bananas are chosen as a main crop, the opportunities for intercropping are excellent (see Figs 69–72). Tall plants such as papaya, castor bean and trees, which grow almost as tall as bananas and have a similar lifespan, can be grown in between the bananas or at the edges of fields. Shorter plants like chilli, cotton and eggplant have shorter lifespans than banana and can be planted thickly between bananas that are not yet fully grown. They can be harvested before the bananas are big enough to shade them excessively.

Maps of intercropping patterns indicate clearly how many of each plant a farmer puts in a field. Figs 68 to 71, inclusive, focus on one small piece of land to show how a small landowner obtains extra yield from the land and keeps enough root systems in place to control net loss of soil fertility through leaching.

Fig. 68 shows typical field patterns and sizes of farm fields in the Girvi area. The village is at the top of the mapped area, with the main road to the village on the right and the main river bed on the left of the map. North is at the bottom of the map. Many of the field boundaries marked on base maps a hundred years ago can still be found and are shown as heavier lines in Fig. 68. Many additional field boundaries have been added since, as the fields are divided into smaller and smaller parcels. Fig. 69 is a more detailed view of the long banana–cotton field running from north to south, near the bottom of Fig. 68.

Fig. 69 shows how a cotton crop was grown in the inter-banana spaces without reducing the resources available for the main crop. Border crops such as soybean and papaya are common and important in both India and Nepal. Intercrops such as onions and coriander, which are much smaller and shorter-lived than cotton or

82 Intercropping

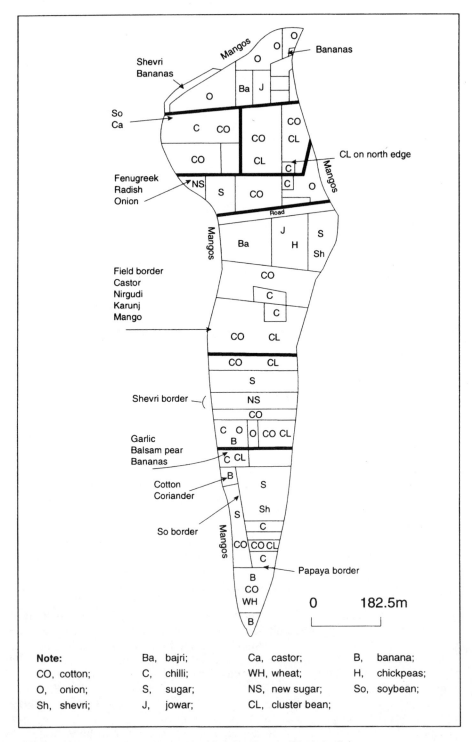

Fig. 68: Map of field patterns in an irrigated area (Girvi, India)

Intercropping in India 83

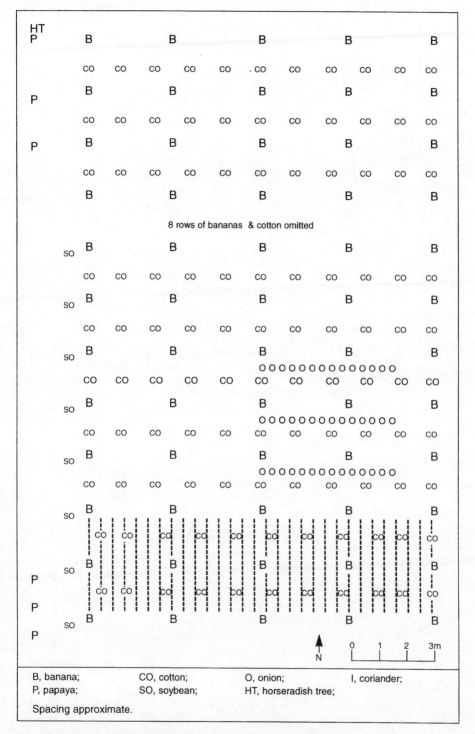

Fig. 69: Map of intercropping in a banana garden (Girvi, India)

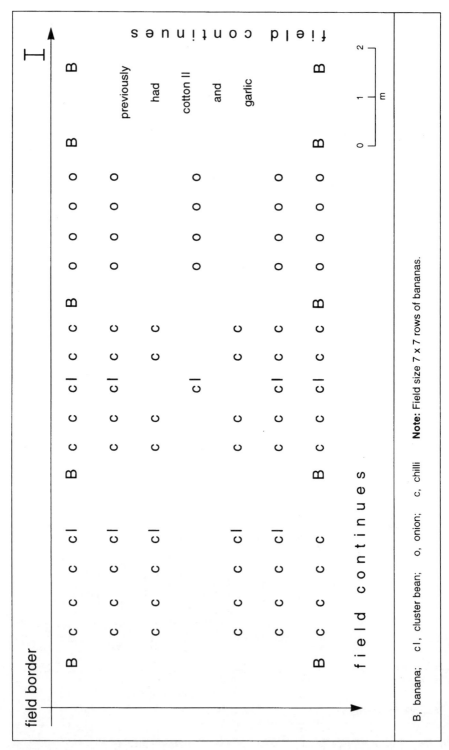

Fig. 70: Detailed map of intercropping in a banana garden (Girvi, India)

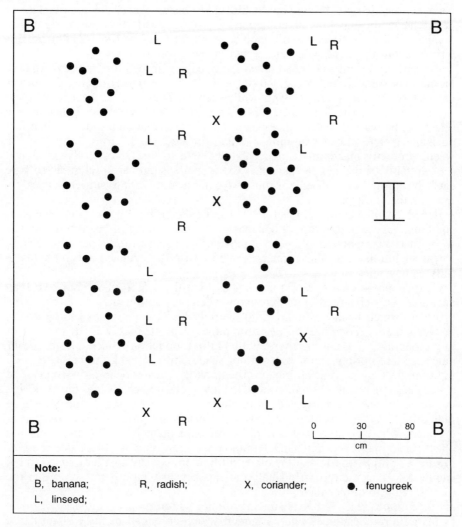

Fig. 71: More detailed map of intercropping in a banana garden (Girvi, India)

banana, can easily be grown with the cotton, just as the cotton is grown with banana.

Fig. 70 shows an even more detailed picture of part of a field, the field directly above the Fig. 69 area. Fig. 69 maps cotton, coriander and banana, but Figs 70 and 71 show that the situation is actually much more complicated. The field data used to construct the graphs of Figs 72 and 73 include all the plants recorded on these maps. The cluster beans shown on Fig. 70 are left over from the time when they were an intercrop of cotton. In India, each cotton and cluster-bean plant is harvested many times in one season as these indeterminate crops continue to produce bolls and beans. In the Fig. 70 map area, chillies and onions were grown. The chillies were probably as old as the cotton and cluster beans, but the onions were planted after the cotton. Of the four squares on this map formed by bananas, the two on the left have chilli and cluster beans, the square right of centre has onions, and the planting pattern of the space farthest to the right is so complex that it is shown on still another map, Fig. 71. The Roman numeral II is written about the same place on the two maps.

A small farmer may not need much fenugreek or linseed for his family, but he or she may want some, as Fig. 71 suggests. By using intercropping, the farmer can get fresh spices for cooking which can be harvested as needed with no expenditure of cash. The farmer is also using many resources which would otherwise be wasted. These spice plants are so small that their measurable effect on banana yield may be nothing. In fact, growing small crops instead of big weeds probably makes more nutrients available to the almost fully-developed root systems of the mature bananas. Extra root systems are preferable to bare earth because they reduce erosion and continuously bring nutrients back up towards the surface.

In Fig. 72, the banana graphs for 1970 and 1980 both show papaya as the most commonly chosen intercrop for bananas, with trees and castor beans important as well. There was much more intercropping with bananas in 1970, perhaps because it was such a dry year that farmers wanted to use the irrigated banana fields as fully as possible.

Chilli peppers (see Fig. 73) are not as long-lived as bananas but live long enough to be well suited to intercropping when they are the main crop. Since the plants are small bushes which are not densely leafy, many crops can be grown between them. As pungency of fruit, rather than size, is valued, the use of intercropping does not interfere with chilli production. Since cluster beans have a single upright stem, they do not shade chillies much but can obtain all the sunlight they need as an intercrop with chilli. Cluster bean grows upward at about the same rate as chilli or cotton; the lower cluster bean leaves die back, and the lower beans are harvested from those parts of the stem which become increasingly shaded, while new leaves and beans are produced near the top of the plant where light is still available. An okra plant is similar in shape to a cluster bean plant, but as it is much taller, only a few okra plants are intercropped in such a chilli field. Most of the other chilli intercrops are smaller than chilli but can be grown as intercrops in small amounts because chilli is not too big or too fast-growing to hinder them.

Since onions (Fig. 72) need sun and are not tall crops, they are usually grown as main crops. Fenugreek and coriander, which are small plants without dense canopies of leaves, can easily be grown as onion intercrops. Eggplants and chillies are bigger than onions and must be widely spaced in onion fields to leave the onions enough resources to produce a worthwhile harvest. If eggplants or chillies were closely spaced in such fields, then one of them would be the main crop instead of onions.

Peanuts (Fig. 72) are seldom monocropped, but give the appearance of so being because so few plants of other species are planted in peanut fields. The tall, thin cluster beans do not hinder peanuts, but soybeans are more competitive unless they are quite widely spaced. Some soybeans were grown as border crops at the edge of peanut fields. Peanut fields cannot accommodate much intercropping.

Cotton (Figs 73 and 74) is a widely-spaced, slow-growing plant suitable for intercropping in its early months of growth. No field observations of early-season intercropping with cotton were made near Girvi; however, the literature

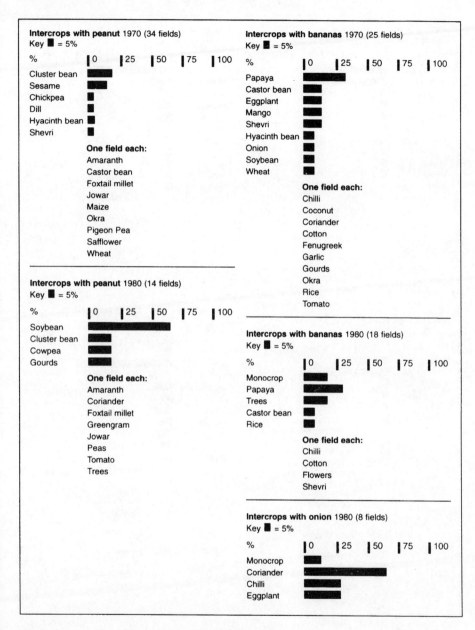

Fig. 72: Peanut, banana, onion intercrops (Girvi, India)

mentions many useful food crops which are often grown with cotton in India. In September, only a few examples could be found of the many intercrops which probably grew in the cotton fields earlier in the year.

Maize (Fig. 74) was not a popular crop in the Girvi area. The dry air and low rainfall make this region much more suited to jowar (sorghum) and bajri (pearl

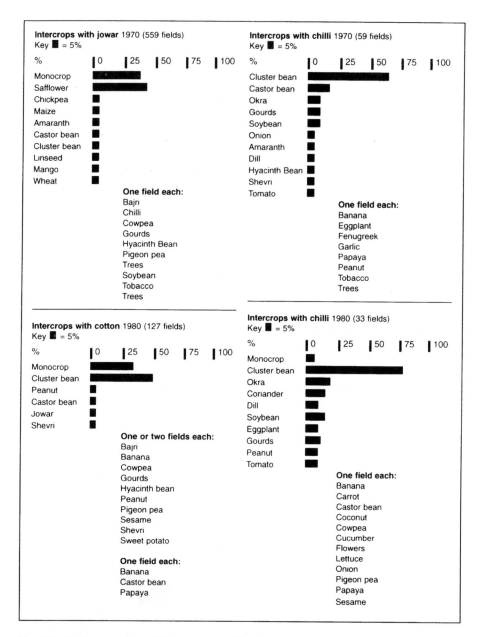

Fig. 73: Jowar, cotton, chilli intercrops (Girvi, India)

millet); likewise with rice (Fig. 74). Where irrigation water is available, farmers find sugar a much more profitable crop than either maize or rice. In 1970, there were fourteen rice fields with five intercrops (coconut, hyacinth bean, sesame, shevri, trees), each of which occurred once only. In 1980, maize and soybeans were quite important rice intercrops. Gourds occurred twice. The crops which

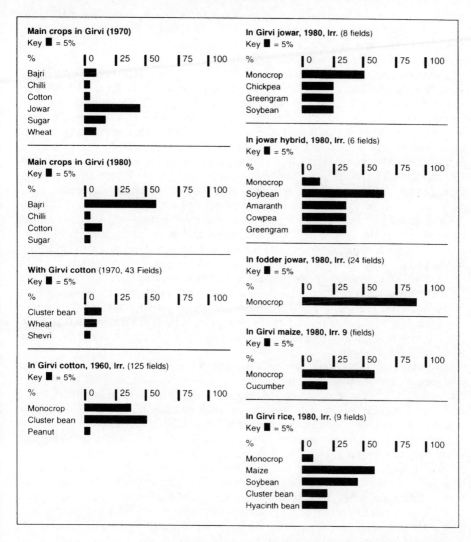

Fig. 74: Main crops in Girvi, 1970 and 1980 – Cotton, jowar, maize, rice intercrops

occurred only once were coriander, cowpea, sesame and moth bean. These were grown on the ridges of the irrigated rice fields to avoid the excessively wet conditions of the main part of the fields.

Jowar was a far more important main crop in Girvi in 1970 than in 1980 because 1970 was a very dry year. The monsoons were so late that few farmers were able to plant the usual bajri crop. When some rain did fall towards the middle of summer, farmers tried to gain some yield by planting jowar, since it was then too late and too dry to plant bajri. In fact, the jowar grew poorly, never produced seed heads, and was used for grazing animals. A very high proportion of the 1970 jowar fields were neither irrigated nor intercropped. Jowar is a tough crop which can endure spells of drought without dying better than can bajri. If the rains had developed in a satisfactory way after the early failure of the 1970 monsoons, more of the jowar could have produced grain. When the rain turned out to be inadequate for grain on non-irrigated fields, the jowar could still be used for fodder and grazing, keeping animals alive which would otherwise have died on the dessiccated, non-cultivated hills.

The most common jowar intercrop in the dry, non-irrigated fields of 1970 was safflower. Safflower is a very spiny small bush which dries up as the oil seeds ripen. It grows quite well in drought conditions, and its spines help keep wandering cattle out of jowar fields. The farmer who is growing the crops may harvest either the safflower or jowar as needed, so his or her cattle, sheep or goats can benefit. A factor which could be relevant here is the practice by big landowners in some parts of India of planting a token crop on some of their land to reduce social pressure to sell or rent the land. In 1970, there were large acreages of fields some distance from the village where jowar plants were very small and no attempt was made at intercropping.

On a few 1970 fields, one or two emergency irrigations were possible in an attempt to save the crop. These irrigated fields had more intercropping. In such a difficult year, the farmers tried to make the best use they could of their best fields. About 5 per cent of the fields had one or more of eight different crops, planted in an attempt to grow needed intercrops which could not be obtained from the failed or unplanted bajri fields. If fields with only one or two specimens of an intercrop are included, there were 18 intercrops with jowar. It is hard to compare this with jowar intercropping in a year of normal rainfall such as 1980, since the farmers tended not to grow jowar as a main crop when the early rains were adequate for bajri.

Jowar is much more intercropped when irrigation is planned than when it is an emergency, last-minute operation. The extra water makes several extra intercrops possible, because jowar is a fairly long-lived crop compared to bajri. In the first three months of growth, jowar has still not attained its full height, so significant amounts of water and nutrients are available for other crops.

Both hybrid and local varieties of jowar are planted with more intercropping when they are irrigated (see Fig. 74). Except for amaranth, all of the intercrops are legumes. A third variety of jowar is used for fodder. Farmers with dairy cows or water buffaloes grew irrigated fodder jowar (hundi) which has long stalks and leaves. Since a number of these stalks are harvested each day, maturity and grain production is not important for this variety of jowar. It performs the same role as silage for a North American dairy farmer, but since this part of India has no winter, fodder can be grown all year if water is available.

Sugar is an irrigated crop in this part of the country. Most sugar fields appeared to be monocropped. Farmers who were asked about intercropping with sugar said that it was not done. But through field observations, it was found that 22 different intercrops were growing with sugar. One explanation of this apparent contradiction is that sugar is quite a small plant for at least the first three months of its year-long growth cycle. During these months, small farmers can grow many food crops as intercrops, but big farmers, who do not need to grow crops for food, tend not to intercrop even when the sugar is small. After three or four months from the time of planting, sugar completely shades the ground and little intercropping is possible. Thus, for most of the time that sugar is in the field, the fields are monocropped except for the presence of a small leguminous tree called shevri. About half of the mature sugar fields had this small, woody tree as an intercrop. It probably helps protect the cane from destruction by the wind, and since it grows at the same rate as sugar, the two do not shade each other too much. Castor bean is a tall, woody plant often grown along the edges of sugar fields. All the other crops shown in the 1970 and 1980 sugar fields (see Fig. 75) are smaller plants which can quickly use the sunlight, water and nutrients that the sugar is not yet able to use, and which can therefore be grown without reducing sugar yields as we have seen in Chapter 1. It must be kept in mind that nutrients, together with water, are always being leached downward through the soil; thus, many of the nutrients used by the intercrops would not be available later for the sugar, even if the intercrops were not grown.

The graphs in Fig. 76 are an attempt to show how one intercrop affects others in sugar fields. The bottom section of graph 1A shows the percentage occurrence of intercrops in all fields where both shevri and sugar are present. The lower part of graph 1B shows the importance of intercrops in fields where both chickpea and sugar were planted. Comparing 1A and 1B shows that there is much less intercropping when shevri is present than when chickpea is present. This is due to the role of shevri, discussed in the preceding paragraph. Castor bean (graph 1C), like shevri, tends not to co-occur with other intercrops. Maize, on the other hand, like chickpeas, is found with other quick-growing intercrops. Sugar, maize and chickpeas together are quite common in young sugar fields in this part of India. Graph 1E shows that coriander and onions do well together as sugar intercrops, just as they did in fields where onions were the main crop.

Bajri is smaller than jowar and has a shorter lifespan. The irrigated three-month variety of bajri has few intercrops, whereas the five-month cultivar has more.

The advantage of growing both short- and long-term varieties of bajri is that the farm family can replenish its grain supply three months from the beginning

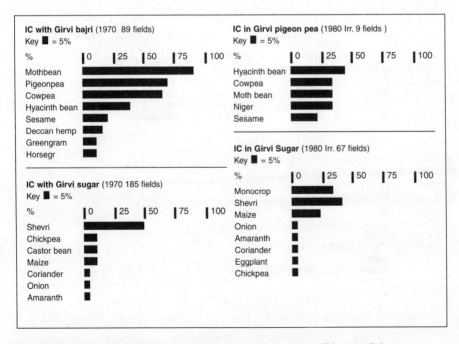

Fig. 75: Bajri (pearl millet), pigeon pea, sugar intercrops (Girvi, India)

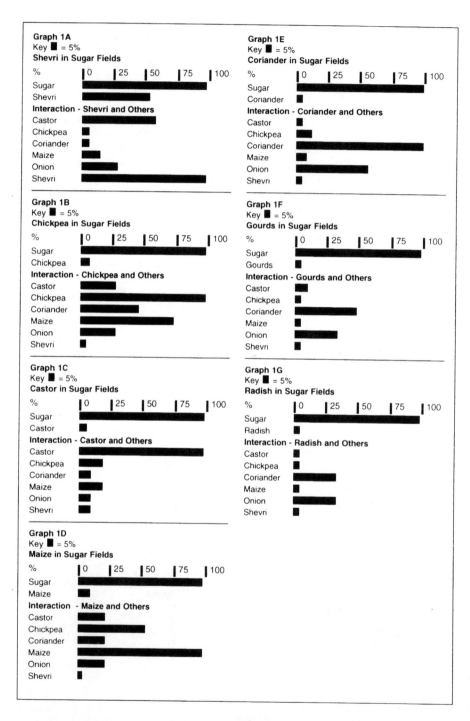

Fig. 76: The effects of one intercrop on other intercrops in sugar fields (Girvi, India)

of the monsoon rains instead of only after five months. Many farmers had divided bajri fields, in which part of the field matured in three months and the other part in five months. The three-month variety usually had two intercrops, cowpea and greengram. Additional intercrops, found less frequently in the three-month bajri fields, were maize and soybeans. Only four of the 96 three-month fields observed were not irrigated. The average five-month bajri field had four intercrops: cowpea, hyacinth bean, moth bean and pigeon pea. Additional intercrops in the five-month bajri fields were Deccan hemp, greengram, horsegram, mustard and sesame. Only two of 64 five-month fields were irrigated. These three-month and five-month bajri fields were included in the following more general study of bajri intercropping.

In 1970, bajri near Girvi was well intercropped with legumes but not, as we shall see, as completely intercropped as the bajri fields on the two Palwan plateaux south of the village. Since it is only the fields near the village that can be irrigated, this is the only place where irrigated and non-irrigated bajri fields could be compared. When intercropping at all three levels is compared, it seems clear that intercropping became more intensive with increasing elevation and distance from the village. Near the village, three-month bajri tends to be irrigated and has less intercropping, while five-month bajri tends to be unirrigated and has more intercropping.

Palwan I

In both 1970 and 1980, the Palwan I plateau, at 3500 feet, was relatively inaccessible by road. The people who lived there had to walk about four miles each way to carry their produce to the market in Girvi and bring home their purchases. This probably provided an incentive to grow as much as they could of what they needed by intercropping.

Very similar intercrops were grown with the bajri fields in 1970 and 1980 (see Fig. 77). As we have seen, in 1970 bajri could not be grown at the 2000-foot level because of the drought, but at the 3500-foot level, the higher, cooler, somewhat wetter environment did make it possible to grow a bajri crop that year. In both 1970 and 1980, all the five most common bajri intercrops were legumes. Diversity of legumes provides different leaf and root configuration in both time and space as well as better protection from insects, disease and weeds; it provides alternative possibilities for whatever weather may occur, and it spreads out the harvesting workload and food yield over a period of several months.

Several other intercrops were grown with bajri. Deccan hemp is a fibre and oil crop with edible young leaves. Niger and sesame are oil crops. Pigeon pea is a seven-month legume which uses sunlight and residual soil moisture long after all the other bajri intercrops are gone. Amaranth is a grain and a spinach, and mustard is a spinach and a spice. With a field of intercropped bajri, a family can obtain a balanced diet and maintain soil fertility indefinitely. For this to be achieved, there must be a balance between the rate at which nutrients are removed from the soil and the rate at which nutrients are returned or made available by formation of new soil from the bedrock.

There was very little jowar on Palwan I in 1970 (four fields) because bajri was grown instead. In 1980, both jowar and bajri fields were common, and both were heavily intercropped. The 1980 jowar crop was almost as heavily intercropped as the bajri crop.

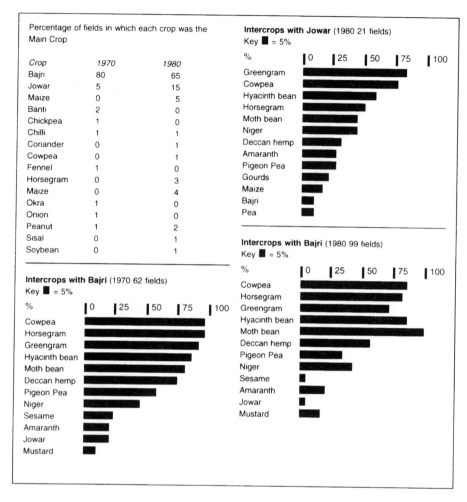

Fig. 77: Bajri, jowar intercrops (Palwan I plateau, India, 1970 and 1980)

When a large enough sample of fields can be found, it is possible to analyse the interaction between intercrops. Fig. 78 is a table which shows the percentage co-occurrence of 11 pairs of intercrops grown with bajri at the Palwan I level in 1970. Column 2 gives the percentage frequency with which each crop occurred in all of the bajri fields. Each of columns 3 to 13 is devoted to one of the 11 intercrops. Column 3, for example, includes figures for all the bajri fields where amaranth was present (13 fields). Deccan hemp occurred in 92 per cent of these amaranth–bajri fields but in only 71 per cent of all bajri fields (column 2). Thus, amaranth and Deccan hemp appear to be compatible crops. These two plants are more or less the same size, so one does not shade the other unduly. Niger and sesame are also similar in height to amaranth and occurred more frequently when amaranth was present. On the other hand, moth bean, the smallest of the crops listed here, occurred less frequently when amaranth was present.

1	2	3	4	5	6	7	8	9	10	11	12	13
	% in all 62 bajri fields	Amaranth (13)*	Cowpea (59)	Deccan hemp (44)	Greengram (57)	Horsegram (57)	Hyacinth bean (54)	Jowar (13)	Moth bean (50)	Niger (28)	Pigeon pea (35)	Sesame (17)
Amaranth	21	100	22	27	19	21	24	23	14	36	23	41
Cowpea	95	100	100	95	96	96	96	100	96	100	94	94
Deccan hemp	71	92	71	100	72	70	80	62	74	86	71	71
Greengram	92	85	93	93	100	93	94	100	92	93	91	94
Horsegram	94	92	93	91	93	100	93	85	94	89	88	82
Hyacinth bean	85	100	86	95	88	88	100	92	88	93	97	94
Jowar	21	23	22	18	23	19	22	100	16	29	31	41
Moth bean	81	54	81	84	81	82	83	62	100	71	77	59
Niger	44	77	46	52	44	44	48	62	38	100	54	41
Pigeon pea	55	62	54	54	54	54	63	85	52	68	100	76
Sesame	26	54	25	25	26	25	30	54	18	25	37	100

*number of fields.

Fig. 78: Intercrop interaction in bajri fields (Palwan I plateau, India, 3500 Feet, 1970)

Several other tall crops also discouraged moth bean. In the jowar, niger and sesame columns (9, 11 and 13), it can be seen that moth bean was chosen less frequently as an intercrop when these three crops were present. This can be seen by comparing these columns with column 2.

Medium-to-small plants – cowpea, greengram and horsegram (columns 4, 6, and 7) – did not seem to affect the presence of other crops.

The presence of taller crops seemed to provide an opportunity for choosing other tall crops for intercropping combinations in the bajri fields. In fields with pigeon pea (a tall but slow-growing crop), the taller crops, hyacinth bean, jowar and sesame, occurred more often. Likewise, in sesame–bajri fields, the tall crops – amaranth, hyacinth bean, pigeon pea and sesame – occurred more often, and the smallest crops, moth bean and horsegram, were chosen less often.

Jowar, the tallest of the crops, encouraged the presence of the tall crops, hyacinth bean, niger, pigeon pea and sesame. The presence of jowar decreased the frequency of Deccan hemp, horsegram and moth bean. Niger, a medium-sized bushy crop, occurred more frequently with Deccan hemp, hyacinth bean, jowar and pigeon peas and less with horsegram and moth bean.

In 1980 (Fig. 79), there was less intercropping in bajri fields on the Palwan I plateau. Since the rains were better, there was less need to intercrop densely to make the best use of water, and the additional rain would make individual plants grow bigger and cause more competition for any given density of intercropping. Cowpea, Deccan hemp, greengram, horsegram, pigeon pea and sesame were chosen less frequently as bajri intercrops in 1980 than in 1970. The Maharashtra guaranteed employment programme could be another possible influence. In this programme, the state provides jobs near an unemployed person's home, such as

1	2	3	4	5	6	7	8	9	10	11	12	13
	% in all 99 bajri fields	Amaranth (18)*	Cowpea (82)	Deccan hemp (59)	Greengram (71)	Horsegram (77)	Hyacinth bean (85)	Jowar (4)	Moth bean (92)	Niger (97)	Pigeon pea (37)	Sesame (5)
Amaranth	18	100	18	17	15	17	18	25	18	21	21	20
Cowpea	82	83	100	85	79	81	86	75	85	85	86	100
Deccan hemp	59	50	59	100	62	60	66	0	61	70	81	100
Greengram	71	61	68	75	100	74	69	50	71	64	54	80
Horsegram	77	72	75	78	80	100	80	75	77	74	78	80
Hyacinth bean	85	89	89	95	83	88	100	75	88	93	100	100
Jowar	4	11	4	3	3	4	2	100	4	2	8	0
Moth bean	92	94	94	97	90	94	95	100	100	94	100	80
Niger	47	56	47	54	46	44	53	25	47	100	57	60
Pigeon pea	37	44	39	49	30	38	43	75	40	43	100	40
Sesame	5	6	6	8	6	4	6	0	4	6	5	100

*number of fields.

Fig. 79: Intercrop interaction in bajri fields (Palwan I plateau, India, 3500 feet, 1980)

building roads, dams or embankments. As a result, the small farmers perhaps see less need to try to maximize yields from any one field.

The effects of intercrops on other intercrops in bajri fields were very similar in 1970 and 1980. In 1980, cowpea, greengram and horsegram, as in 1970, had no effect on the frequency of other crops. Moth bean also had no effect; perhaps because with less intercropping in 1980, moth bean did not have to be omitted from the crop combinations. As in 1970, the tall crops tended to be grown more frequently with other tall crops.

Palwan II

On Palwan II, at the 4000-foot elevation plateau, as on Palwan I at the 3500-foot level, there was much more intercropping than at the Girvi level in both 1970 and 1980 (see Figs 80 and 81). The most striking change in the ten-year period is the decline in the number of bajri fields on Palwan II. During that time, a road was built on the side away from Girvi. It is possible that many small farmers were influenced by the new access road to change their main crop from bajri to coriander; however, it seems more likely that big landowners controlled this choice of crops. The marked change from food crops to a commercial crop would seem to be a choice that big farmers would make. Many small farmers seem to have been deprived of owned or rented land, and some of them are being hired to grow a commercial crop with only a little intercropping. Farmers grew as many as 15 crops in one field on Palwan II in 1970, but in 1980 many of the same people were employed in building a road to give access to the plateaux from the Girvi side.

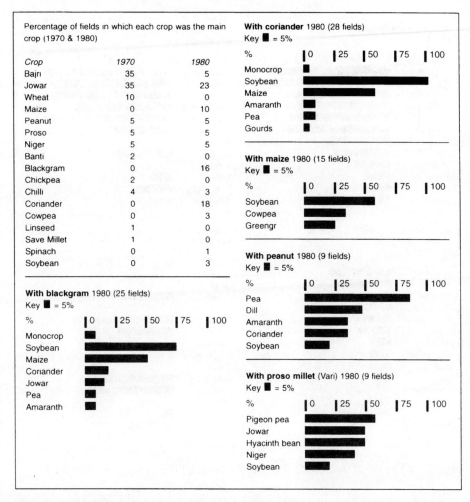

Fig. 80: Blackgram, coriander, maize, peanut, proso millet intercrops (Palwan II plateau, India)

Coriander, blackgram and maize were not used as main crops on Palwan II in 1970 but were important in 1980 (see Fig. 80). Widely-spaced soybeans and maize were the most important intercrops for blackgram and coriander but were unimportant in 1970. Maize, soybean, cowpea and greengram were intercrops which are used when considerable soil moisture is available, as in the irrigated fields near Girvi. There is no irrigation on the Palwan II plateau which is at the top of the mountain, but the horizontal layers of Deccan lavas trap rainwater close to the soil surface if the rains are good. Soil water drains slowly towards the edge of the plateau and was observed in a couple of rice fields by the edge of a cliff, where the water-table and the soil surface intersected. There are wells in the middle of the Palwan II plateau where the water-table in a wet year can be observed not far below the surface.

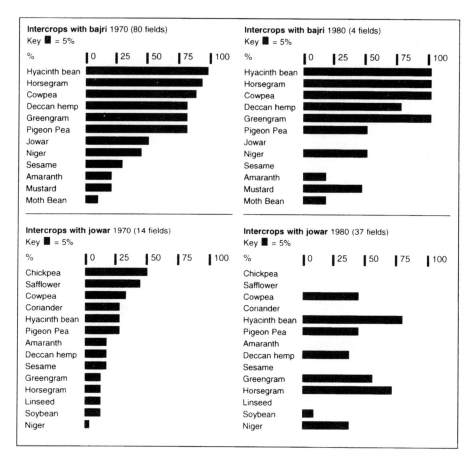

Fig. 81: Bajri, jowar intercrops (Palwan II plateau, India, 1970 and 1980)

Peanut, proso millet, bajri and jowar were main crops in a few fields in both 1970 and 1980 on the plateau. Presumably, these are fields still controlled by small farmers. There was a good deal of intercropping in these fields even though proso millet and peanut are short, sun-loving crops. Coriander, peas and soybeans were intercropped with peanuts in 1980 but not in 1970. Only one new intercrop was grown with proso millet in 1980 that was not grown in 1970 – soybean. It is possible that these were attempts to try out new intercrops, or it could be that the rains were so much better in 1980 that crops needing more water could be intercropped.

Jowar intercropping on Palwan II involved fewer different intercrops in 1980 but more frequent use of the crops that were chosen. In 1970, coriander was sparsely sown as a bajri intercrop, but in 1980, coriander was an important main crop. Drought-resistant safflower was grown in the dry 1970 season but not in the 1980 season of normal rainfall.

Figs 82 and 83 show two extremes of bajri intercropping. At the 2000-foot Girvi level, the market is nearby, some bajri fields are irrigated, and there are more big landowners, all of which seem to reduce the tendency to intercrop. At

1	2	3	4	5	6	7	8	9	10	11	12	13	14
	% in all 501 bajri fields	Amaranth (2)*	Cowpea (275)	Deccan hemp (39)	Greengram (226)	Horsegram (43)	Hyacinth bean (103)	Jowar (7)	Moth bean (240)	Niger (10)	Pigeon pea (121)	Sesame (33)	Soybean (30)
Amaranth	–	–	1	–	–	–	1	–	–	–	1	–	–
Cowpea	55	–	100	97	63	67	88	71	65	100	87	64	40
Deccan hemp	8	–	14	100	1	2	35	–	15	20	16	21	–
Greengram	45	–	51	5	100	44	14	86	17	30	23	18	80
Horsegram	9	–	10	3	9	100	7	–	17	30	17	9	–
Hyacinth bean	21	–	33	89	7	14	100	–	41	40	63	55	–
Jowar	1	–	1	–	3	–	–	100	–	–	–	–	7
Moth bean	48	–	57	97	18	98	95	–	100	100	98	82	–
Niger	–	–	3	5	1	5	4	–	3	100	4	3	–
Pigeon pea	24	–	39	54	13	49	76	–	50	60	100	39	–
Sesame	7	–	8	21	3	9	18	–	13	20	12	100	–
Soybean	6	–	4	–	9	–	–	29	–	–	–	–	100

*number of fields.

Fig. 82: Intercrop interaction in bajri fields (Girvi, India, 2000 feet, 1980).

the 4000-foot Palwan II elevation in 1970, there was no easy access to market, no irrigation, and small farmers had more control over the crops grown. Comparing columns 2 of the two figures shows how much more intercropping there was on Palwan II in 1970 than in Girvi in 1980.

Cowpeas and greengram were popular at both levels. At the Girvi level, the most frequent combinations included cowpea, greengram and moth bean in about half the fields, and in many cases they all occurred together. Pigeon pea and hyacinth bean occurred in a quarter of the bajri fields and were usually present with cowpea, greengram and moth bean. Of the other crops, Deccan hemp was more common when hyacinth bean, niger, pigeon pea and sesame were grown and less common with greengram, horsegram, jowar and soybean. Deccan hemp seems to do best with crops of about the same size and is chosen less often in the presence of crops that are both smaller, such as greengram and horsegram, or taller, such as jowar and soybean. Many observations can be made from Fig. 82. For instance, moth bean, a small plant, did not do well with greengram but grew very frequently underneath Deccan hemp, hyacinth bean, pigeon pea and sesame. It also seemed compatible with horsegram, the other small legume. At the 4000-foot elevation, however, moth bean was almost completely displaced by horsegram. It is possible that when the ecological niche underneath all the other crops, occupied by both moth bean and horsegram, becomes a little damper and shadier, the horsegram with its fine, thin leaves is a better crop than the leathery-leafed moth bean. Pigeon pea, the tallest of the legumes, grew more frequently when other taller crops such as Deccan hemp, hyacinth bean, niger

1	2	3	4	5	6	7	8	9	10	11	12	13
	% in all 80 bajri fields	Amaranth (19)*	Cowpea (68)	Deccan hemp (64)	Greengram (65)	Horsegram (71)	Hyacinth bean (76)	Jowar (19)	Moth bean (9)	Niger (34)	Pigeon pea (64)	Sesame (25)
Amaranth	24	100	24	23	26	24	25	53	22	9	30	27
Cowpea	85	84	100	84	91	87	89	79	100	85	88	77
Deccan hemp	81	79	79	100	78	82	82	79	89	82	81	88
Greengram	81	89	87	80	100	87	84	79	78	76	80	92
Horsegram	90	89	93	91	97	100	93	89	78	88	89	92
Hyacinth bean	95	100	100	97	98	99	100	95	100	91	97	100
Jowar	24	53	22	23	23	24	24	100	22	26	27	31
Moth bean	11	11	13	11	11	10	12	11	100	12	13	12
Niger	43	16	43	44	40	41	41	47	44	100	41	38
Pigeon pea	79	100	82	81	78	79	82	89	89	76	100	85
Sesame	31	37	29	36	37	34	34	42	33	29	34	100
Soybean	–	–	–	–	–	–	–	–	–	–	–	–

*number of fields.

Fig. 83: Intercrop interaction in bajri fields (Palwan II plateau, India, 4000 feet, 1970)

and sesame were present. Pigeon peas also did well with horsegram and moth bean, perhaps because the pigeon peas grow so slowly.

At the 4000-foot level in 1970, bajri intercropping was at its most sophisticated (Fig. 83). It is clear from the figures in column 2 that most fields had six or seven crops. Six intercrops grew in 79 per cent or more of the bajri fields. Cowpea, Deccan hemp, greengram, horsegram, hyacinth bean and pigeon pea seemed to grow well together and with the other intercrops. These six intercrops grown with bajri would seem to be a basic and successful combination at this level. With this combination, a local micro-climate was created in the bajri fields which conserved moisture by raising the humidity among the densely-planted crops. This would reduce water loss through transpiration for all the crops.

The other intercrops of Fig. 83 were sometimes chosen to accompany the basic seven crops and sometimes not. Here, too, some farmers seemed to intercrop more than others. Jowar appeared in only a quarter of the fields but seemed to grow particularly well with amaranth. Amaranth also occurred in a quarter of the fields and did well in jowar–bajri and poorly in niger – bajri fields. Sesame was present in about a third of all the bajri fields, and niger was found in about 40 per cent of the fields except when amaranth was present.

An attempt to analyse Figs 78, 79, 82 and 83 by shortening and simplifying them, as was done with the data in Figs 31, 35 and 41, did not seem possible. Unlike the Jamaican study area, where some of the crops are perennials, all the crops grown with bajri in this dry part of India are annuals. The smallest crops (moth bean and horsegram) are not the fastest-growing plants. The dry beans of moth bean and horsegram are harvested after four to five months, so their ecological niche is not that of small crops grown between immature plants of longer-lasting crops. They are not, therefore, like the kidney beans of Jamaica, which are pulled out before the larger crops get too big. Instead, these small beans provide the lowest level of sun-catching leaves throughout the lifespan of most of the larger crops in the field.

The crops harvested first (cowpea after two-and-a-half to three months and greengram after three to four months) are bigger than moth bean and horsegram. The removal of these 30–40cm tall bush beans certainly leaves more water, light and nutrients for other crops, including smaller ones, but makes construction of a graph with the smallest, first-harvested crops on one side, and the tallest later-harvested crops on the other, impossible.

Another problem encountered in trying to explain this Deccan agriculture in its simplest terms is that legume pods can either be harvested green for a vegetable or dry as a grain. Assigning a harvest date for analytical purposes is therefore difficult. The same problem applies to mustard and amaranth which can supply leaves for vegetables early in the season and/or seeds later on.

Still another factor to be considered is that many of the crops are available to the farmer in varieties which ripen earlier or later. For example, the harvest dates of bajri or jowar can vary by a couple of months. The oil seeds (sesame and niger) ripen in four to five months as do Deccan hemp, amaranth and mustard. Some crops do not need to be harvested immediately when they are ripe, so the farmer can choose many different harvest dates to avoid having too much work come due at the same time. Bajri, the main crop, does have a lot of plants ripening at the same date, so even small farmers must hire extra labour. The bullrush-shaped heads of bajri contain completely unprotected seeds which make them very susceptible to bird damage. It is, therefore, important to harvest all the bajri as soon as it is ripe.

By the time all the crops listed in Figs 78, 79, 82 and 83 (except pigeon pea, a seven-month crop) are harvested, the pigeon pea has grown into quite a large deep-rooted bush. Leaves of one pigeon-pea plant may even touch those of another, using sunlight, shading and cooling the ground but not hindering any other crop in the last two months of its growth because all others have been harvested.

The intercropping practices observed in the Girvi area suggest that this is a highly sophisticated, reliable method of farming. Variations in planting combinations over time and from place to place are evidence of constant fine-tuning in which large and small farmers can respond quickly when opportunities either become available or disappear. The effects of some of the changes that are occurring on the maintenance of soil quality will be discussed in the next chapter.

5 The earth's soil and its future

INTERCROPPING CAN MAINTAIN soil fertility, monocropping cannot, where all other factors are equal. The fertility of the world's soils was built up by a great variety of wild plants. There were usually many species growing at once on any given hectare of land. Traditional agriculture using intercropping may not build up soil fertility any further, but where suitable plant combinations are used, it can maintain soil quality at a given level more or less indefinitely. Monocropping obtains high yields by using mineral fertilizers to supply some elements and by depleting soils to obtain other elements.

In temperate lands where many organic-matter decay vectors are depleted each winter, natural soil fertility builds up as generation after generation of vegetation dies and adds organic matter to the upper layers of weathered rock. Over periods of millions of years, plants and plant associations have evolved which fit into different climate and environmental niches. Where one species left a good deal of the light, water and soil nutrients unused, another species would evolve to make use of these resources. The bedrock supplies the soil minerals necessary for soil formation, but bedrock, even weathered bedrock, is not soil, and farmers cannot farm weathered bedrock successfully. Plants use their root systems to recover the soil minerals they need, so when plants decay, they enrich the soil with exactly the minerals required by the plants, plus organic matter containing nutrients derived and manufactured from elements in the air, water and earth. Under natural vegetation, in temperate lands, organic matter is created faster than decay vectors can destroy it, forming more and more soil as time goes on. And in temperate lands, organic matter decays slowly, making all the necessary soil nutrients gradually available more or less at the rates required by new generations of plants.

In a natural ecosystem, there are many different plants holding their leaves at different levels and angles to catch solar energy. Since all the plants together use a great deal more solar energy than any one species could use, a plant association will produce more organic matter (or biomass) per hectare than would be the case if only one species of plant was growing there.

Similarly, plants with different root systems, maturing at different times of year will use a higher proportion of the available water than one species could use.

The same factors apply to soil nutrients. In a natural ecosystem, very few nutrients are permanently removed from the soil. All the organic matter produced by plant growth is returned to the soil. If organic matter is added to the soil faster than decay, erosion and leaching can remove it, then the soil will become richer and richer. This, as we have seen, is how soils are formed and are built up. When modern agricultural practices began in Iowa, some farmers were able to grow crops for 50 years before they had to start adding mineral fertilizer.

The presence of many root systems in the soil makes it possible to produce a lot of biomass in a year, but it also serves to prevent the products of organic decay from escaping the soil layer too rapidly. The root systems at different depths and densities recycle nutrients and keep the soil rich. Without all these intertwining natural root systems, the constantly downward-leaching rainwater in most parts of the world would remove nutrients from the soil faster than they could be created or recovered.

In tropical regions there is no periodic set-back to the population of decay vectors in the soil. Where there is no winter, the fungi, bacteria and soil creatures which carry decay-producing organisms from one place to another are always present in sufficient numbers to produce rapid decay of anything that falls to the ground. Fruits, leaves, branches and animal manure decay quickly, and the breakdown of large molecules makes nutrients available to other plants. Tropical jungles survive by having multitudes of root systems which are able to absorb nutrients as soon as decay makes them available. If the plants do not immediately trap the downward-leaching nutrients, they will soon leach below the level of root systems and then down to the water-table, the rivers and the sea. Tropical soils are generally very low in nutrients and organic matter. Clearing the forest for cultivation allows cropping for one, two or three years until the organic matter is all rotted and leached away. Yields then become so low that it is not worth while to farm the soil.

In tribal areas of Thailand, areas cleared of trees to make crops possible for two or three years were observed. The hill people leave high stumps so that the original jungle trees will remain alive and capable of reclaiming the agricultural land, converting it back into jungle so that soil fertility can be rebuilt. Some modern farm experts, products of a culture which believes that history proceeds in stages, believe that these tree stumps are the result of very primitive efforts to clear the land (Richards, 1985). If the wild vegetation is killed on purpose or through bad techniques and is not able to recover, the soil rapidly becomes so poor that trees are no longer able to grow there. This is the explanation for the expanding areas of Imperator grass in the Philippines, Africa and other tropical countries. This grass makes few nutritional demands on the soil and contains so few nutrients that animals cannot live on it. Such grasses grow on poor soil and do not have deep root systems which could build up the soil. They deny use of the land to farmers and grazing animals.

Intercropping and soil maintenance

Traditional farmers must maintain soil fertility or give up farming. It is a basic fact of life, which Carl Sauer pointed out very clearly, that any society which destroys the resources on which it depends cannot survive for very long. Traditional agriculture with intercropping is superior to monocropping in this respect because it has been developed and tested over a very long period. Many traditional societies developed farming systems which preserved soil in usable condition for hundreds or even thousands of years. Other ancient societies, such as some of those which developed cities, destroyed their soils causing their societies to collapse. Still other civilizations such as China and Japan, which consciously cared for their soils, survived.

Since intercropped fields have more leaves over a longer period of time than monocropped fields, they use a higher proportion of the available solar energy. As a result, more organic matter is produced per field, and there is a better chance of maintaining soil fertility. At the same time there are higher total yields of food crops for the farmer (see Appendix 1).

Intercropping uses rainwater or irrigation water to produce more growth than most monocrops could produce, where all other factors are equal. Also, intercropping with its many root systems utilizes and retrieves nutrients better than monocropping.

When intercropping is carried out systematically, it can support people indefinitely. For any given soil and intercropping system, there is a population of human beings which can be maintained indefinitely. Many traditional societies developed philosophies of life, lifestyles and standards of living which did not demand more of the soil than it could produce on a sustained basis. Unsustainable rates of resource use may produce much food for a short time but deplete the productive capacity of agricultural soils. Perhaps spiritual, artistic and social progress can be improved indefinitely, but material growth must be limited to what is materially possible.

Bad agriculture drives out good

A brief history of the relationship between the development of modern agriculture and the deterioration of soil might read somewhat as follows.

In the Middle Ages in Europe, people still practised intercropping and planted mixed seeds by broadcasting (scattering them by hand). The old nursery rhyme, 'Oats, peas, beans and barley grow' probably gives a good idea of one kind of medieval crop combination. Each plant was harvested with a sickle as it became ripe.

In 1740, Jethro Tull, the English agricultural innovator, had completed most of his inventions. His achievement was to make agriculture more profitable. He advocated planting only one crop per field, planting in rows, and using horse-drawn machines to cultivate between the rows. This increased the productivity of those farm workers who retained their jobs. It also reduced the number of workers, reduced labour costs for big landowners, and increased profits. And it made surplus population available for emigration overseas. When these migrants arrived in their new homelands, they tended to replace native intercropping with the new monocropping. Jethro Tull and the big landowners claimed that yields were increased by the new methods. However, as we have seen in previous chapters, this is true only in the sense that one crop per field usually gives a higher yield of that crop than the yield of that same crop in a field where it shares resources with intercrops. In fact, the total yield in an intercropped field is almost always higher than the yield of one crop, where all other factors are equal (see Appendix 1). Unfair and unscientific propaganda for modern monocropping is still used by proponents of big profits and big farms to favour short-term profits in all parts of the world.

By 1840, monocropping had reduced the fertility of European and other monocropped soils. In that year, Justus von Liebig published the suggestion that nitrates, phosphates and potash be added to soils. By 1840, too, European scientists knew enough about chemistry to know some of the elements a plant needed. They discovered that shortage of N, P and K were limiting plant growth. The monocropping which Jethro Tull introduced and popularised had allowed the organic content of the soil to be reduced. It was found profitable to obtain N, P and K from mines and add them to the fields as needed. Where lack of N, P and K had been the limiting factors, their addition led to greater plant growth, greater yields and greater profits. The long-term effects on soils and mineral deposits were not considered important. The influence of Adam Smith, Ricardo and the 'invisible hand' of economics, which would make this the best of all possible worlds if each person maximized his or her short-term profits, was very strong at this time. Mines for mineral fertilizers were not yet very deep, ores were still rich, and the energy to mine them was obtained cheaply from coal.

By 1940, N, P and K were being added as a routine modern farm practice in countries around the world. The addition of these three important elements made it possible to flush out from the soil all the other elements plants need. Where plant growth is limited by lack of N, P or K, plants are not able to reduce to dangerous levels the availability of the ten or so trace elements which soils contain and which plants also need. Trace elements did not become scarce until N, P and K were added to soils. By 1940, some farms which had depended heavily on N, P and K for some time were starting to suffer from the depletion of one or more trace elements such as boron, calcium, copper, iron, magnesium, manganese, molybdenum, selenium, sulphur and zinc. A high-enough level of organic matter in soil can, of course, supply the needed trace elements as well as improve soil structure and water-holding capacity. The addition of N, P and K makes organic matter and trace elements become limiting factors and rapidly reduces their presence in the soil. Fukuoka speaks of mineral fertilizers as 'burning out' the soil.

By the 1980s, in all parts of the world, there were many farms which were no longer profitable. There were only a few places like China and relict indigenous cultures (the Fourth World) where methods were retained which farmed labour-intensively, using intercropping and related practices to build up organic matter and maintain soil quality. As prices of energy to mine and transport minerals tend to rise, and as mines get deeper and poorer in quality, more and more farmers find modern agriculture unprofitable. Ever-larger machines are introduced to try to reduce costs so that monocropping practices can be continued for a few more years. In the Palouse district of the state of Washington, for example, machines which cultivated a swath 50-feet wide are being replaced with tractors and cultivators enabling one worker to cultivate strips 90-feet wide. Farmers without sufficient resources lose their land to those who are able to buy the new machines.

In some communist and socialist countries, where many people are given responsibility for small private plots of land, high yields are achieved. In China and Russia some thought is being given to assigning more responsibility to individual small farmers. But most communist countries depend on modern machine agriculture just as private capitalism does. In Western Europe and America, there is some intercropping in community gardens and allotments. In the United States, many people of African, Vietnamese and Latin American origin practise intercropping and would be able to make productive use of more land if some way could be developed to give them access to land.

Research on intercropping

The types of controlled research described in Chapter 1 need to be continued, expanded and made more sympathetic to peasant knowledge. And more field research needs to be carried out to understand the success of intercropping and other traditional agricultural methods. One method of field research described in previous chapters involves listing the crop combination of each field. When lists for several hundred fields have been compiled, a computer can be used to determine what secondary crops are grown most often with maize, which are grown with potatoes, and so on. Maps showing every plant in a section of one field can be used to show how many plants of each type the farmer grows and how closely he spaces them. The work of travelling over the world and recording such traditional crop combinations for various environments and types of plants is only just beginning. It is important for research workers to comprehend the expertise of the traditional small farmer. Asking questions about planting methods and yields in the language of the small farmer will establish communication and can lead to important discoveries from the vast store-house of traditional knowledge. It is, apparently, the knowledge we need to save the world's soils.

6 Social and economic implications of agricultural practices

LITERATURE WHICH FAVOURS large-scale modern agriculture tends to claim that if land were returned to traditional farmers, then millions of people would starve. Traditional farmers, when they are mentioned at all, are presented in college textbooks, newspapers, news magazines and comic books as very set in their ways, unable or unwilling to respond to new ideas or opportunities. The author's experience has been quite the opposite. In fact, peasant farmers grow much more from given resources of land, water and sunlight and make better use of mineral fertilizers if they are available, than do large-scale farmers. They are observant and innovative when it comes to developing or accepting new plant material or other things which are really helpful. They tend not to accept poor ideas or poor plant material, and modern innovators who are responsible for such ideas retaliate by claiming that the peasants are hide-bound in their ways.

Current publications propagate two sorts of false arguments which lead to the unwise removal of skilled traditional farmers from the land: first, they compare the yield of a monocrop with the yield of the same crop in an intercropped field; second, they compare the yield of a fertilised monocrop with the yield of an unfertilized intercrop. In a fair comparison, where equal amounts of nutrients and environmental resources are available for both methods of planting and where the yield of all the intercrops plus their main crop is compared with monocropping, intercropping produces the greater yield. As Appendix 1 shows, some intercropping combinations, for which ancient agriculturalists selected and bred the crops to be grown together, give two or three times the yield for the same resources.

According to the modern ideas of economists and businessmen, success means only a return on investment, so 'success' can be achieved even if a farming area is rendered unusable for many years after a business venture has left the area. When soils are left diseased, leached out, laterized, salinized, water-logged or eroded away, or when the water-table is lowered by pumping so that poor people cannot reach it, the agribusinessman is apparently not really concerned for the future of the soil or of humanity. Such concerns are sometimes even called 'non-economic side-effects'. Does this mean that some economists and some scientists have tacitly decided that the earth's people cannot afford to have a future?

In a comparison of intercropping and monocropping, one can say that intercropping tends to maximize total crop yields and soil fertility, whereas mechanized monocropping tends to maximize only financial profit for a few landowners.

When a landlord hires labour to do the work in his fields, he often finds that the fourth, fifth or sixth intercrop (or even the second and third) cost more in labour than they are worth when sold. This is partly because growing crops for sale to poor people tends to be uneconomic since the poor lack money; consequently, prices for these staple crops tend to be low. If the landowner rents out his land with the agreement that he will get half the harvest of the main crop, he may believe (probably rightly) that the food intercrops are reducing the yield of the main cash crop and so try to discourage intercropping. Until recently, the scarcity of research on intercropping allowed the landowner to believe that intercropping is not efficient. Land is so scarce in countries like India that a big landowner can easily choose tenants who will comply with this point of view.

It should be understood that poor people are not necessarily the ones who suffer most when business conditions in the First World are bad. In several Latin American countries, good prices for export crops often lead to plantation expansion, landowners keeping poor people off unused acreages and Indians being driven from the land. When prices of export crops are low and sales are poor, Indian squatters (on land they once owned but were never paid for) are more tolerated and are able to feed their families by intercropping.

No one can advocate the collapse of modern world economic systems as a solution to world economic problems. However, in some ways it could be said that the quickest means to economic recovery for most Third World people might be the collapse or partial collapse of the First World.

As Frances Moore Lappé points out in *Aid As Obstacle* and other studies, the aid money which is sent to poor countries usually goes into plantation development to promote the growth of export crops. Such large numbers of small farmers are displaced by modern development that many poor countries now import not only N, P, K, tractors, fuel and insecticides, but also food. The resulting imbalance of payments and loan repayment problems have become highly complex and apparently insoluble.

Food produced in the United States by using the cheap minerals and fertilizers which are temporarily available from a historical point of view and by using large machines which reduce labour costs is often so cheap that food prices in poor countries are reduced to the point where local food-producing farmers go bankrupt. The local farmers who can no longer put in the work needed for intercropping, terracing, grassed waterways, cover crops, green manure and practising staggered phenology lose land which then becomes available for monocropped plantation export crops.

Why do aid programmes not succeed in helping very many people? It seems that the overriding economic policies and attitudes of First World peoples toward Third World peoples are so much more powerful than aid programmes aimed at the really poor that such aid cannot accomplish much. Let us briefly examine some of the economic forces at work.

Free-enterprise corporations set up plantations in many Third World countries. Each corporation, in co-operation with a few local people who become a relatively wealthy élite, tries to produce export crops (for the First World) as cheaply as possible. It buys or acquires land as cheaply as possible, does nothing to help set up labour unions, and does little to improve the social infrastructure (education, health care, housing, etc.) for most people in the Third World country. Keeping the wages and living standards of most Third World countries low has been a conscious policy of big business since Ricardo's time in the 1820s. What is new about the twentieth century is that an excessive number of plantations overproduce export products. The world market for sugar, coffee, cotton, tea, rubber, jute, chocolate, palm oil, coconut oil and many other products cannot consume all that is produced. So the prices fall, which further impoverishes the Third World. But the low prices and currency devaluation benefit the people of the First World, so they become richer. These simple effective mechanisms continue to drain wealth from poor countries and benefit wealthy countries. Aid programmes accomplish little in reversing this great flow of economic wealth.

The debt system set up by First World monetary systems and banks is another aspect of the problem. Modern banks are much like loan sharks. For poor people and poor countries, also, this can be called debt peonage. When interest rates on the debt rise, as happened in the early 1980s, it is even more difficult for Third World countries to pay the interest on the debt, let alone to pay off the principal. When interest payments are not made, the International Monetary Fund, controlled by First World countries, arranges the devaluation of Third World currencies, which makes exports even cheaper for First World countries, while further impoverishing the poor. As prices of exports drop, some poor countries realize that all the money derived from exports is equal to their interest on their loans. They have no money to buy gasoline, buses, food, machetes or other manufactured goods (Kwitny, 1984).

Where can new facts and ideas be obtained which will break this vicious circle? As we have seen, the poor people of Third World countries have some important facts and ideas at their disposal. The rich élites of these poor countries tend to be too modernized to understand the value of traditional agriculture. We have seen that Third World peasants intercrop because it gives higher total yields, protects and even builds up the soil, provides more employment in growing food, keeps insects, weeds and plant diseases under control, and can even produce some export produce in the same fields where food is grown. If all tropical farmland was in small intercropped farms (like those in Taiwan), there would not be as much export produce as there would be with monocropping, but the advantage of this is that surpluses would be reduced and the price of export crops increased. If Third World countries had more money, and all Third World individuals had some money, there would be a greater market for First World manufactured goods, and more payment on debts might be made. First and Second World people need to be educated to accept this and learn to be happier with fewer material things.

A world system which would work, and which needs to be set up, not just for social justice but to keep the rest of the earth's soil from sliding into the ocean, might look something like this. Taiwan and the rest of China are already accomplishing much of the following. The soil needs to be looked after by many more supervisors than are currently used in most countries. The proprietary and territorial instincts of humans need to be harnessed so that small farmers will plant their crops carefully. They must be people who care about the world, the future and their families and not just about short-term profits. Perhaps co-operatives or land trusts or state ownership of the land is needed to keep wealthy people from buying up the poor people's land. The small farmers must learn to intercrop if they have forgotten how. Research teams can explore regions where modern agriculture has not yet destroyed traditional techniques. Not only planting patterns in space and time need to be noted, but plant species and varieties most suited to intercropping need to be saved as well. The research teams which are looking for valuable genetic variations in Third World crop species could extend their work to search out the cultivars best suited to various intercropping combinations.

Using intercropping and the accumulated traditional knowledge of many centuries, numerous supervisors who care about the future can feed their families, keep their families employed, keep soil on the land, utilize more of the available sun and water than monocropping can use, and solve many cultural and agricultural problems. It is probable that most of the world's people can be, and need to be, involved in this effort. Planting patterns and other techniques which have evolved because they worked successfully at making continued agriculture possible can work just as successfully in the future. Everyone must learn what peasant farmers have learned – that only the usufruct can be consumed if we want to live on the earth indefinitely. On a small planet, with a limited amount of soil (created by plants over a period of thousands of years), it is only possible to harvest a certain amount each year. Since modern agriculture depends on non-renewable resources which are becoming increasingly scarce, it must be abandoned (or modified) in favour of a continuing kind of agriculture. Individual enterprise, personal responsibility, small farms and co-operatives have most important contributions to make towards a viable future for the world's people.

Much of the blame for Third World problems which modern society usually assigns to the population explosion could more accurately be assigned to the deprivation of people who have had their land taken away from them, quite often by development projects. The displaced families often move to the cities where they try to earn money to buy the food that they can no longer grow. However, their city diet is much less varied and can lead to malnutrition. The families who have been driven off their carefully farmed lands by plantations, dam building and flooding are suddenly defined as a population problem. Does this mean that they should have had foreknowledge of large-scale commercial developments and not had the children to whom they hoped to pass on their fertile fields?

Peasant farmers, whose goal is maximum possible total yield for the resources available, are not easily influenced by the criticism that intercropping might reduce the yield of the main crop. All over the Third World, proponents of Green Revolution methods have been puzzled by peasant lack of interest in monocropping. Perhaps because methods being used by small farmers have not been understood, aid providers have given up their attempts to help small farmers and have helped big farmers by default. But as we have seen, the traditional agricultural practices observed in Jamaica, India, Nepal and other parts of the world provide excellent evidence that small farmers make effective use of the land. Much excellent evidence now exists for arguments maintaining that small farmers should have the land.

Gene Wilken's book, *Good Farmers* (1987), gives many excellent examples of the ways in which Mexican and Central American traditional farmers manage their resources. Many traditional small farmers produce food for their own families and grow commercial crops at the same time in the same field. They carefully classify their soil according to their own system (often quite unaware that Mexican soil scientists have classified the same soils in modern ways). The traditional farmers have developed methods of soil treatment (such as adding organic matter, green manuring, applying silt from rivers and lakes, terracing, intercropping and crop rotation) which keep up the productivity of the soil. Where necessary, they have developed hand- or animal-powered methods for irrigating or draining the land. Gene Wilken is a geographer who has done many years of field work in Central America.

F.H. King's book, *Farmers of Forty Centuries* (1911), analyses traditional Chinese agriculture. Since King had been head of the US Soil Survey, his enthusiasm for traditional methods should affect the thinking of those who are trying to understand the best ways to keep up the productivity of the world's soils. Most traditional agriculture produces far more food per unit of available plant nutrient than does modern agriculture. In addition to many other techniques such as composting, the Chinese have always recognized the importance of returning the nutrients in human waste to the soil. There are scholars who feel that the persistence of Chinese civilization for 4000 years without a collapse is at least partly due to Chinese understanding of the soil and its needs.

Paul Richards is an anthropologist who has been studying West African traditional agriculture. His book, *Indigenous Agriculture Revolution* (1985), contains many wonderful examples (examples to be wondered at) of modern methods which have done a great deal of harm to the people and soils of West Africa. In one example, modern methods were introduced on an emergency basis during World War II but proved so much less productive than traditional methods that the old ways had to be re-introduced even before the war was over. In addition to these books, many research articles on traditional agriculture are now being published.

Why have scientists been so slow in beginning to recognize the achievements of traditional farmers? Apart from a certain cultural hegemony from which European peoples benefit, it seems likely that basic philosophic attitudes are involved here. Repeatedly in the intercropping literature there are statements that indigenous agriculture is too complicated to be analysed scientifically. And it is difficult to imagine scientists, who are trained to test painstakingly the influence of one factor at a time, being enthusiastic about a holistic system like intercropping which integrates a dozen factors at once to produce a highly productive, socially advantageous, non-destructive type of agriculture. When scientists propose to increase the yield per acre of one particular crop for the good of humanity, the critic who questions these accomplishments by pointing to increased soil erosion, dependence on non-renewable resources, unemployment and world hunger may be accused of being non-scientific. The scientist who is under contract to increase monocropped yields may succeed in achieving the stated goal and, at the same time, may label terrible resulting problems as 'side-effects', which cannot be considered because they are beyond the scope of the contracted research. One goal of the present book is to assist in providing a scientific vocabulary for the systems of agriculture that traditional farmers have developed so that they can be tested for their scientific validity (Goonatilake).

It will require the efforts of many people to work out the details of a change-over to an agriculture which has a stable future and is able to preserve soil quality. As it becomes clearer what needs to be done, it seems likely that great improvements in soil quality and human welfare may be possible. Among other problems, it needs to be realized that public discussion of a new long-term agriculture is not likely to be supported by the news media, which are in part paid for by short-term profits obtained by depleting the resources of the future.

First World people need to point out to Third World élites that the ordinary people of those countries have effective methods for solving many or all current food and agricultural problems, if they were allowed to have good land. Too often in the twentieth century, modern agriculture and aid programmes in the Third World have tended to take over the best land and drive traditional farmers into urban slums or mountain hillsides. Since the hungry poor do not have the immediate resources to build hill terraces, disastrous deforestation and erosion can occur. The local élite and educated people of the modern world then say, and write in their journals and newspapers, that traditional farmers do not know how to farm and should not be allowed to have the land.

Modern mechanized farmers now expect heavy subsidies from other taxpayers to keep in business a type of agriculture which has no long-term future. To a large extent, it is left to small farmers and non-farm concerned citizens to work towards a new type of agriculture that is sustainable over the long term.

Interesting new ideas for future agriculture are being developed. Wes Jackson in Kansas is developing perennial soil-holding crops, and Mollison in Australia is developing tree-crop combinations. The author of this book hopes to add intercropping and the vast accomplishments of traditional farmers to the methods being considered for solving world-hunger problems, unemployment problems and the need for a permanent type of agriculture.

When schools, universities and the unbiased public have become sufficiently aware of the agricultural achievements of Third World farmers, it will become possible to help instead of hinder those who know how to care for the world's soils. One advantage of turning to intercropping to solve agricultural problems is that millions of farm experts already exist.

The first step in rebuilding world agriculture after two centuries or so of destruction (for short-term profits) is to stop driving traditional farmers off the land. The second logical step would seem to be to get them back on the land. In Europe, small farms are protected and encouraged. In parts of Sweden, when a small farm comes up for sale, only other small farmers are allowed to bid for the land in order to keep land speculators and agribusiness from owning more land. Nebraska has passed Land Trust laws for the same purpose.

It will be necessary for many people to realize the benefits of small-scale farming allowing intercropping and careful soil maintenance before big landowners will willingly give or sell their land for use as small family farms. When farms in the United States and Canada are seized for non-payment of taxes, the current state of public understanding causes this farmland to be auctioned off to big farmers or non-farming landowners. If the general public understood the situation, land seized by the government could be used to set up small farms and provide employment in profitable, ecologically-efficient, non-subsidized agricultural methods. Bankrupt farmers should be offered a part of the land they once owned. Mennonite family farms, which average 60 acres, could provide one successful model for future farms. Many poor people from city slums may not want to participate in such an enterprise, but a start could be made with those who are enthusiastic. As they succeed, others might become interested.

Appendix I: Land equivalent ratios

This appendix is a compilation of the results of several hundred experiments which have been done on intercropping. Overall, intercropping (growing two or more crops together in a field) has been found to use resources more efficiently, and to produce higher combined yields than monocropping (growing one crop per field).

Comparisons are made using the Land Equivalent Ratio (LER), which expresses intercropped yields as a ratio of monocropped yields. For instance, in an intercropped field where the total yield gives an LER of 1.42, the ratio of intercropped yields to monocropped yields is 1.42: 1. This means that, all other factors being equal, the crops when grown together produced 42 per cent more yield than the same crops when grown separately, as monocrops. Particularly compatible crops (bred for compatibility over a long period) have a yield two or three times greater when intercropped than when monocropped.

The LER for an intercropped field is calculated by adding together the LERs of each of the crops. The LER for a single crop in a crop combination is called a partial LER. For instance, if beans intercropped with corn have a partial LER of 0.8, this means that the yield of the beans when intercropped was 0.8 times (or 80 per cent of) the yield of beans grown as a monocrop, all other conditions being equal. If the partial LER of corn in this case was 0.62, then the total LER would be 0.8 + 0.62 = 1.42. In the Appendix, partial LERs are shown in parentheses following the crop name.

Another measure employed in the Appendix is the financial land equivalent ratio (fLER). The LER is based on crop weight or volume; the fLER is based on monetary yield. For instance, a financial LER of 2.00 means that the total monetary yield from intercropping is twice that of monocropping.

The experimental results given in the Appendix are grouped according to main crop. Within each grouping, the experiments are numbered alphabetically by author, giving the author's name and the year of publication of the reference. The author's name can be used to refer to the Bibliography which contains the complete references. (Note that pl/ha stands for plants per hectare.)

In the listing of each experiment, the place where the research was done is indicated after the author's name. This is followed by an alphabetical list of the intercrops used in the experiment, followed by a list of the variables tested. Where indicated, particular parameters of the experiment are explained. Results of the experiment are then given, in terms of weight of crop yields, LER and fLER. If the original author(s) did not calculate LERs or fLERs, but provided the data from which one or both of these ratios could be calculated, this has been done and these figures are included.

Due to the difficulties of experimenting with several crops at once, most of the experiments summarized here deal with only two crops at a time. Special attention should be paid to the experiments, marked with an asterisk (*), which combine three or more crops, since these more closely approximate the results obtainable by intercropping as it is practised in the field.

Bajri

1. Chowdhury, S.L. (1979), Bijapur, India
Intercrops used: Pigeon pea

Results:	LER	fLER
Bajri 970kg/ha (0.87) and Pigeon pea 1670kg/ha (0.61)	1.48	1.31
Bajri 1410kg/ha (0.81) and Pigeon pea 1700kg/ha (0.47)	1.28	1.12

2. Chowdhury, S.L. (1979), Sholapur, India
Intercrops used: Pigeon pea

Results:	LER	fLER
Bajri 1670kg/ha (0.99) and Pigeon pea 2380kg/ha (0.63)	1.62	1.42
Bajri 1930kg/ha (1.04) and Pigeon pea 1970kg/ha (0.96)	2.00	1.96

3. Chowdhury, S.L. (1979), Rajkot, India
Intercrops used: Greengram

Results:	LER	fLER
Bajri 770kg/ha (0.89) and Greengram 540kg/ha (0.11)	1.00	0.88
Bajri 880kg/ha (0.68) and Greengram 570kg/ha (0.21)	0.89	0.83

4. Raikhelkar, S.V. and U.C. Upadhyay (1980), Maharashtra, India
Intercrops used: Blackgram, Cowpea, Greengram, Peanut, Pearl millet
Variables tested: proportion of intercrops

Results:	LER	fLER
MC Bajri (paired rows) 1490kg/ha		
MC Bajri (paired rows 30cm and 70cm) 1835 Rs/ha		
Pearl millet single row 50cm	0.98	0.97
Paired rows Bajri (1.19) and 1 row Blackgram	1.35	1.35
Paired rows Bajri (1.07) and 2 rows Blackgram	1.26	1.29
Paired rows Bajri (1.09) and 1 row Cowpea	1.23	1.22
Paired rows Bajri (1.01) and 2 rows Cowpea	1.20	1.20
Paired rows Bajri (1.13) and 1 row Greengram	1.29	1.21
Paired rows Bajri (1.04) and 2 rows Greengram	1.23	1.22
Paired rows Bajri (1.12) and 1 row Peanut	1.27	1.23
Paired rows Bajri (1.07) and 2 rows Peanut	1.22	1.27
Paired rows Bajri (0.70) and 1 row Bajri	0.97	0.96
Paired rows Bajri (0.58) and 2 rows Bajri	0.89	0.90

5. Reddy, M.S. and R.W. Willey (1979), Hyderabad, India
Intercrops used: Peanut

Results:	LER
MC Bajri 1227kg/ha; MC Peanut 1185kg/ha	
Bajri (0.55) and Peanut (0.71)	1.26

6. Singh, P. and N.L. Joshi (1980), Rajasthan, India
Intercrops used: Cluster bean, Cowpea, Dewgram, Greengram, Sesame
Variables tested: proportion of intercrops
Experimental parameters: N, P and K were added; Bajri was grown in single and double rows; results are an average of 2 years, 1976 and 1977

Results:	LER 1 row	LER 2 rows	fLER 1 row	fLER 2 rows
Bajri and Cluster bean	1.14	1.21	1.18	1.21
Bajri and Cowpea	0.94	0.91	1.03	1.09

Bajri and Dewgram	1.30	1.54	1.22	1.40
Bajri and Greengram	1.09	1.19	1.41	1.72
Bajri and Sesame	1.00	1.31	1.07	1.46

1 row Bajri and 1 row Greengram   1.06   1.64

7. Singh, S.P. *et al.* (1979), New Delhi, India
Intercrops used: Greengram
Variables tested: planting pattern, timing
Results: *LER*
MC Bajri planted 9 July 2457kg/ha; MC Greengram 699kg/ha
Paired rows:
Both seeded 9 July: Bajri (1.00) and Greengram (0.15)   1.15
Both seeded 29 July: Bajri (0.85) and Greengram (0.36)   1.21
Bajri transplanted 29 July (1.05) and seeded Greengram (0.37)   1.42
Both seeded 18 Aug.: Bajri (0.34) and Greengram (0.45)   0.79
Triple Rows:
Both seeded 9 July: Bajri (1.09) and Greengram (0.28)   1.37
Both seeded 29 July: Bajri (0.90) and Greengram (0.60)   1.50
Bajri transplanted 29 July (0.98) and Greengram (0.71)   1.69
Both seeded 18 Aug.: Bajri (0.32) and Greengram (0.60)   0.92
Bajri transplanted 18 Aug. (0.63) and Greengram (0.93)   1.56

8. Tiwana, M.S. and K.P. Puri (1979), Punjab, India
Intercrops used:
Variables tested: Forage bajri
Experimental parameters:
Results: *LER*
Green fodder:
1973
MC Bajri 1326q/ha
Lucerne   1.13
Berseem   1.40
1974
MC Bajri 1169q/ha
Lucerne   1.38
Berseem   1.42
Dry matter:
1973
MC Bajri 213q/ha
Lucerne   1.14
Berseem   1.42
1974
MC Bajri 190q/ha
Lucerne   1.36
Berseem   1.39

9. Wahua, T.A.T. and D.A. Miller (1978), Illinois, USA
Intercrops used: Greengram
Results: *LER*
Bajri and Greengram   0.94

10. Willey, R.W. (1979a), Hyderabad, India
Intercrops used: Sorghum
Variables tested: millet and sorghum genotypes

Results:	LER
Millet Gam 75 (1.06) and Sorghum GE 196 (0.25)	1.31
Millet Gam 75 (0.61) and Sorghum IS9237 (0.35)	0.96
Millet Gam 75 (0.53) and Sorghum CSH6 (0.53)	1.06
Millet Gam 75 (0.55) and Sorghum Y75 (0.69)	1.24
Millet Gam 73 (0.82) and Sorghum GE 196 (0.22)	1.04
Millet Gam 73 (0.56) and Sorghum IS9237 (0.42)	0.98
Millet Gam 73 (0.51) and Sorghum CSH6 (0.48)	0.99
Millet Gam 73 (0.55) and Sorghum Y75 (0.57)	1.12
Millet PHB 14 (0.85) and Sorghum GE196 (0.20)	1.05
Millet PHB 14 (0.59) and Sorghum IS9237 (0.45)	1.04
Millet PHB 14 (0.54) and Sorghum CHS6 (0.34)	0.88
Millet PHB 14 (0.52) and Sorghum Y75 (0.80)	1.32
Millet Ex-Bornu (1.03) and Sorghum GE196 (0.14)	1.17
Millet Ex-Bornu (0.71) and Sorghum IS9237 (0.39)	1.10
Millet Ex-Bornu (0.72) and Sorghum CSH6 (0.38)	1.10
Millet Ex-Bornu (0.71) and Sorghum Y75 (0.42)	1.13

11. Willey, R.W. and M.S. Reddy (1981), Andhra Pradesh, India
Intercrops used: Peanut
Variables tested: root interaction
Experimental parameters: Bajri roots were separated from Peanut roots by a plastic partition underground

Results:	LER
MC Bajri 2853kg/ha; MC Peanut 1235kg/ha	
Control: Bajri (0.50) and Peanut (0.72)	1.22
Partitioned: Bajri (0.43) and Peanut (0.76)	1.19

Banana

1. Devos, P. and G.F. Wilson (1979), Nigeria
Intercrops used: Cassava, Maize
Variables tested: density, three-crop combination

Results:	LER
MC Banana 17.5tons/ha	
Banana and Maize	1.21
* Banana and Maize and Cassava	1.39
Half the density of Banana, resulted in higher Cassava yield	1.34

Barley

1. Chowdhury, S.L. (1979), Jhansi, India
Intercrops used: Chickpea

Results:	LER	fLER
Barley 2080kg/ha (0.82) and Chickpea 1100kg/ha (0.35)	1.17	1.15

2. Chowdhury, S.L. (1979), Agra, India
Intercrops used: Chickpea, Mustard
Results:

	LER	fLER
Barley 1450kg/ha (1.03) and Chickpea 620kg/ha (0.39)	1.42	2.10
Barley 1400kg/ha (0.73) and Mustard 1990kg/ha (0.34)	1.07	0.90

3. Martin, M.P. and R.W. Snaydon (1982), England
Intercrops used: Bean
Variables tested: density, planting pattern, proportion of intercrops
Results: LER

Experiment I
MC Barley 3.9tons/ha; MC Bean 7tons grain/ha
50–50 mixture – data from graphs

	LER
Mixed in same row: Barley (1.08) and Bean (0.45)	1.53
Alternate rows: Barley (1.07) and Bean (0.78)	1.85

Experiment II
MC Barley 0.9tons/ha; MC Bean 3.5tons/ha

	LER	Average
Mixed in same row:		
75 Barley–25 Bean: Barley (0.9) and Bean (0.2)	1.10	
50 Barley–50 Bean: Barley (0.6) and Bean (0.5)	1.10	1.10
25 Barley–75 Bean: Barley (0.4) and Bean (0.7)	1.10	
Alternate rows:		
75 Barley–25 Bean: Barley (1.1) and Bean (0.4)	1.5	
50 Barley–50 Bean: Barley (1.0) and Bean (0.5)	1.5	1.49
25 Barley–75 Bean: Barley (0.73) and Bean (0.74)	1.47	

Cassava

1. Chew, W.Y. (1979), Malaysia
Intercrops used: Chilli, Longbean, Peanut, Sorghum, Tobacco
Variables tested: timing, two- and three-crop combinations
Results:

	LER
Cassava and chilli – relay	1.98
Cassava and longbean	1.78
Cassava and sorghum	1.53
Cassava and tobacco	1.76
Cassava and chilli	1.52
Cassava and longbean	1.74
Cassava and peanut	1.70
Cassava and tobacco	1.52
* Longbean and cassava and longbean – 2 relays	2.48
* Peanut and cassava and peanut – 2 relays	2.22
* Sorghum and cassava and sorghum – 2 relays	2.01

2. Moreno, R. and R.D. Hart (1979), Costa Rica
Intercrops used: Bean, Maize, Sweet Potato
Variables tested: timing, two- and three-crop combinations
Results:

	LER
Cassava and Bean	1.45
Cassava and Maize	1.40

Cassava and Sweet potato	1.17
Cassava and later planted Sweet potato	1.99
* Cassava and Bean and Maize	2.20
* Cassava and Bean followed by Sweet potato	2.42
* Cassava and Maize followed by Sweet potato	2.45
* Cassava and Sweet potato followed by second crop of Sweet potato	1.95
* Cassava and Bean and Maize, followed by Sweet potato	2.82

3. Porto, M. (1976), Brazil
Intercrops used: Bean, Maize, Rice
Variables tested: two-, three- and four-crop combinations

Results:	LER	
2 IC	1.67	average
* 3 IC	1.65	average
* 4 IC	1.57	average
Cassava and Bean	1.97	
Cassava and Maize	2.75	
Cassava and Rice	1.06	
* Cassava and Bean and Maize	2.09	
* Cassava and Bean and Rice	1.13	
* Cassava and Maize and Rice	1.40	
* Cassava and Bean and Maize and Rice	1.57	

4. Sinthuprama, S. (1979), Thailand
Intercrops used: Cob maize, Grain corn, Mungbean, Peanut, Soybean
Variables tested: proportion of intercrops

Results:	fLER
Cassava and Cob maize	1.06
Cassava and Grain corn	0.95
Cassava and Mungbean	1.32
Cassava and Peanut	1.15
Cassava and Soybean	1.32

	Mungbean LER	*Peanut LER*	*Soybean LER*
Cassava and 1 row	1.46	1.45	1.02
Cassava and 2 rows	1.51	1.60	1.32
Cassava and 3 rows	1.75	1.44	1.20

5. Thung, M. and J.H. Cock (1979), Colombia
Intercrops used: Bean
Variables tested: timing
Experimental parameters: 80-day Bean and 340-day Cassava were used.

Results: Cassava and Bean:	LER
Bean planted 6 weeks before Cassava	1.60
Bean planted 4 weeks before Cassava	1.68
Bean planted 3 weeks before Cassava	1.58
Bean planted 2 weeks before Cassava	1.65
Bean planted 1 week before Cassava	1.73
Bean planted at the same time as Cassava	1.75
Bean planted 1 week after Cassava	1.45
Bean planted 2 weeks after Cassava	1.60

Bean planted 3 weeks after Cassava	1.50
Bean planted 4 weeks after Cassava	1.58
Bean planted 6 weeks after Cassava	1.43

6. Wilson, G.F. and M.O. Adeniran (1976), Nigeria
Intercrops used: French bean, Okra, Tomato
Variables tested: density
Experimental parameters: Tomato was followed by Okra, which was followed by French bean; for the first four results below, French bean spacing was 100cm × 30cm, Okra spacing was 100cm × 30cm and Tomato spacing was 100cm × 60cm; for the fifth result below, spacing for all crops was 100cm × 100cm; Cassava spacing was varied as indicated

Results:	LER
* Cassava (300cm × 100cm) and French bean and Okra and Tomato	1.97
* Cassava (300cm × 50cm) and French bean and Okra and Tomato	1.82
* Cassava (200cm × 100cm) and French bean and Okra and Tomato	2.78
* Cassava (200cm × 100cm) and French bean and Okra and Tomato	2.20
* Cassava (100cm × 100cm) and French bean and Okra and Tomato	1.62

Chickpea

1. Willey, R.W. and M.R. Rao (1981), Andhra Pradesh, India
Intercrops used: Safflower

Results:	LER
Chickpea and Safflower	1.18

Chilli

1. Balasubrahmanyan, R. (1950), Andhra Pradesh, India
Intercrops used: Cotton

Results:	fLER
14 Chilli plants and 1 Cotton plant	1.69

Coconut

1. Ramadasan, K. *et al.* (1978), Malaysia
Intercrops used: Cocoa

Results:	LER	fLER
Ramadasan estimates profitability at moderate yields for Coconut and Cocoa		5.51
Coconut 1600kg/ha – Malay$272; Cocoa 800kg dry leaves Combined net profit Malay$1504, 551 per cent of MC Coconut MC Coconut (copra) 1628kg/ha		
Coconut (copra) and Cocoa together yielded 2210kg/ha A neighbouring Coconut crop without Cocoa decreased 329kg/ha, a 27 per cent decrease.	1.36	

Coffee

1. Chengappa, P.G. and N.S.P. Rebello (1977), Kerala, India
Intercrops used: Arecanut, Banana, Ginger, Orange, Pepper, Pineapple, Soapnut, Turmeric
Experimental parameters: results are from a survey of intercropping on 62 estates

Results:	fLER
Net advantage of intercropping with Coffee	1.11

The intercrops used in order of importance were: Orange, Pepper, Banana, Ginger, Turmeric, Arecanut, Pineapple and Soapnut.

2. Ramakrishnan, N.T.V. (1976), Kerala, India
Intercrops used: Elephant foot yam, Ginger, Turmeric
Experimental parameters: results are an average of 2 years, 1973 and 1974

Results:	fLER
Coffee and Elephant foot yam	2.30
Coffee and Ginger	2.10
Coffee and Turmeric	2.60

Coffee was not yet bearing, but was growing better because of intercropping; there is no yield for coffee until it is 8 or 9 years old.

Cotton

1. Anthony, K.R.M. and S.G. Willimott (1957), Sudan
Intercrops used: Peanut
Variables tested: timing

Results:	LER
MC Cotton seed 846kg/ha	
Cotton and Peanut, both sown same date as MC Cotton	1.50
Cotton sown 14 days later than MC Cotton and Peanut sown same date as MC Cotton	1.53

2. Anthony, K.R.M. and S.G. Willimott (1957), Sudan
Intercrops used: Cowpea, Peanut, *Phaseolus angularis*, Soybean
Variables tested: timing
Experimental parameters: results are from 1952–53.

Results: LERs

	Early Planted Legume		Later Planted Legume	
Cotton planted:	June 2	June 30	June 2	June 30
MC Cotton yield:	1516kg/ha	1133kg/ha	1668kg/ha	1172kg/ha
Cotton and Cowpea	1.37	1.83	0.97	1.40
Cotton and Peanut	1.88	2.54	1.52	1.89
Cotton and *Phaseolus angularis*	1.63	2.08	1.33	1.70
Cotton and Soybean	1.27	1.40	1.02	1.27

The farther apart the planting dates are, the higher the intercropped yield.

Bean should be planted in April for best results.
Early cotton gives best monocropped yields.
Later cotton (with greater separation of dates to maturity) gives higher LERs.

3. Arangzeb, S.N. (1966), Bangladesh
Intercrops used: Hill rice, Maize, Paddy rice
Variables tested: cotton productivity, soil type

Experiment I
Experimental parameters: results obtained in Chittagong Hills, with good cotton yields.

Results:		fLER
MC Cotton net profit 109Rs/acre	(5 year average)	
Good Cotton yields in line	Rs87/acre	0.80
Cotton and Paddy rice – broadcast	Rs70/acre	0.64

Experiment II
Experimental parameters: results obtained in red laterite soil, with poor cotton yields.

Results:		fLER
MC Cotton	Rs44/acre	
Cotton and Hill rice	Rs111/acre	2.52
Cotton and Maize	Rs106/acre	2.40

4. Braud, M. and F. Richez (1964), Central African Republic
Intercrops used: Maize, Peanut

Results:	LER	fLER
1st year – Cotton and Maize	1.80	1.83
2nd year – Cotton and Maize	1.90	1.89
1st year – Cotton and Peanut	1.89	2.20
2nd year – Cotton and Peanut	1.93	1.86

5. Carvalho, M. (1969), Mozambique
Intercrops used: Sisal

Results:	LER
Cotton and Sisal, approx.	2.00

6. DeVotta, A.D. and S.R. Chowdappan (1975), Madras, India
Intercrops used: Coriander, Foxtail Millet, Greengram
Variables tested: proportion of intercrops

Results:	fLER
MC Cotton, normal planting, net profit Rs1813/ha	
MC Cotton, paired rows	0.81
1 row Cotton and 1 row Coriander	0.71
1 row Cotton and 1 row Foxtail Millet	0.82
1 row Cotton and 1 row Greengram	1.47
2 rows Cotton and 1 row Coriander	0.74
2 rows Cotton and 1 row Foxtail Millet	0.94
2 rows Cotton and 1 row Greengram	1.04
2 rows Cotton and 2 rows Coriander	0.81
2 rows Cotton and 2 rows Foxtail Millet	1.03
2 rows Cotton and 2 rows Greengram	1.14

7. Divekar, C.B. and F.B. Kurtakoti (1961), Karnataka, India
Intercrops used: Peanut
Experimental parameters: results are an average of 9 years
Results: fLER
                        Net return/acre
MC Cotton   Rs220
Cotton and Peanut   Rs355   1.61

8. IRAT #358 (1965), Haute Volta (Burkina Faso)
Intercrops used: Niebe bean
Results: LER
Cotton 1820kg/ha (0.91) and Niebe bean 1565kg/ha (1.99)   2.90

9. Joshi, S.N. and H.U. Joshi (1965), Gujarat, India
Intercrops used: Peanut
Variables tested: proportion of intercrops
Experimental parameters: results are an average of 3 years
Results: LER   fLER

Cotton: Peanut
  1:1 – Cotton 811kg/ha (0.93) and Peanut 641kg/ha (0.47)   1.40   1.29
  1:2 – Cotton 792kg/ha (0.91) and Peanut 1020kg/ha (0.75)   1.66   1.59
  1:3 – Cotton 667kg/ha (0.76) and Peanut 1170kg/ha (0.84)   1.60   1.50

10. Kairon, M.S. and D.S. Nandal (1971), Haryana, India
Intercrops used: Cowpea, Mungbean
Variables tested: proportion of intercrops
Results: LER
MC Cotton 1562kg/ha seed Cotton

Proportion of Cotton: Cowpea
2:1   1.27
1:1   1.29
1:2   1.09

Proportion of Cotton: Maize
2:1   1.17
1:1   1.49
1:2   1.31

11. Kubsad, S.C. and V.S. Dasaraddi (1974), Karnataka, India
Intercrops used: Chilli, Jowar, Onion, Paddy rice
Results: LER   fLER
MC Cotton 1200kg/ha
Cotton and Chilli   1.14   0.86
Cotton and Jowar   1.83   1.14
Cotton and Onion   2.30   1.24
Cotton and Paddy rice   3.00   1.35

12. Patel, P.K. *et al.* (1979), Gujarat, India
Intercrops used: Peanut, Soybean
Variables tested: proportion of intercrops
Results: fLER
MC Cotton gross income Rs2858/ha
Cotton and 2 rows Peanut 1.20
Cotton and 1 row Peanut 1.12
Cotton and 2 rows Soybean 0.89
Cotton and 1 row Soybean 0.90

13. Singh, S., R. Singh and O.S. Tomar (1973), Punjab, India; and Kairon, M.S. (1971), Haryana, India
Intercrops used: Cowpea
Experimental parameters: results are an average of 6 experiments
Results: LER
Cotton and Cowpea 1.18

14. Varma, M.P. and M.S.S.R. Kanke (1969), Bihar, India; Braud, M. and F. Richez (1964), Central African Republic
Intercrops used: Peanut
Experimental parameters: results are an average of 4 years
Results: LER
Cotton and Peanut 2.21

15. Varma, M.P. and M.S.S.R. Kanke (1969), Bihar, India; Singh, S., R. Singh and O.S. Tomar (1973), Punjab, India; and Kairon, M.S. (1971), Haryana, India
Intercrops used: Greengram
Experimental parameters: results are an average of 7 experiments
Results: LER
Cotton and Greengram 1.25

Cowpea

1. Greenland, D.J. (1975), Nigeria
Intercrops used: Maize
Variables tested: soil preparation
Results: fLER
MC Cowpea 1185kg/ha – ploughed and ridged
Cowpea and Maize – ploughed and ridged 2.00
Cowpea and Maize – ploughed and flat 1.81
Cowpea and Maize – strip tillage 2.18
Cowpea and Maize – zero tillage 2.27

Finger millet

1. Chowdhury, S.L. (1979), Bangalore, India
Intercrops used: Green chilli, Lab-Lab

Results:	LER	fLER
Finger millet 3820kg/ha (0.31) and Gr. chilli 320kg/ha (0.62)	0.93	0.67
Finger millet 2380kg/ha (0.35) and Gr. chilli 1110kg/ha (0.63)	0.98	0.88

Finger millet 2690kg/ha (0.72) and Lab-Lab 1310kg/ha (0.35) 1.07 1.08
Finger millet 2360kg/ha (0.72) and Lab-Lab 940kg/ha (0.20) 0.92 0.99

2. Pillai, M.R. et al. (1957), Madras, India
Intercrops used: Cotton, Peanut
Variables tested: two- and three-crop combinations
Experimental parameters: results are an average of 4 years 1951–1955.
Results: LER
MC Finger millet 1797kg/ha; MC Cotton 831kg/ha;
MC Peanut 2355kg/ha
Finger millet (0.94) and Cotton (0.53)   1.47
Finger millet (1.09) and Peanut (0.83)   1.92
Cotton (1.12) and Peanut (0.62) – 1 year only   1.74
* Finger millet (0.95) and Cotton (0.31) and Peanut (0.59)   1.85

3. Pillai, M.R. et al. (1957), Madras, India
Intercrops used: Cotton, Peanut
Variables tested: two- and three-crop combinations
Results: LER   fLER
MC Finger millet 1290kg/ha; MC Cotton 254kg/ha;
MC Peanut 1459kg/ha
Finger millet (0.64) and Cotton (0.72)   1.36   1.34
Finger millet (0.88) and Peanut (0.80)   1.68   1.63
* Finger millet (0.83) and Cotton (0.72) and Peanut (0.65)   2.20   2.16

Garden crops

1. Cox, J. (1979), Pennsylvania, USA
Intercrops used: Bean, Bush bean, Cabbage, Carrot, Green bean, Lettuce, Maize, Pepper, Soybean, Tomato
Variables tested: two-crop combinations
Experimental parameters: results are from garden situations, with no main crop
Results: LER
Bean (1.00) and Cabbage (1.23)   2.23
Bean shade enabled Lettuce to thrive in summer   1.70
Bush bean (1.00) and Carrot (0.50)   1.50
Green bean (1.54) and Pepper (0.64)   2.18
Maize (1.00) and Soybean (0.35)   1.35
Soybean (0.66) and Tomato (1.00)   1.66

2. Norman, D.W. (1974a), Nigeria
Intercrops used: Cotton, Cowpea, Millet, Peanut, Sorghum, Sweet potato
Variables tested: labour costs, two-, three- and four-crop combinations
Experimental parameters: there is no main crop; results are averages
Results:

	Labour not costed	All labour costed	June and July labour costed
Net Return ($/acre): MC Cotton; MC Peanut; MC Sorghum	20.8	10.4	18.7

Appendix I: Land equivalent ratios

	fLER	fLER	fLER
2 *crops*: Cotton and Cowpea; Millet and Sorghum; Peanut and Sorghum	1.57	1.60	1.53
3 *crops*: * Cotton and Cowpea and Sweet Potato * Cowpea and Millet and Sorghum * Millet and Peanut and Sorghum	1.48	1.50	1.39
4 *crops*: * Cowpea and Millet and Peanut and Sorghum	2.14	2.60	2.05

Grass

1. Daulay, H.S. (1978), Rajasthan, India
Intercrops used: Cluster bean, Cowpea, Dewgram, Greengram
Experimental parameters: *Cenchrus ciliaris* grass was used;
  total dry matter was compared with dry matter from grass alone

Results:	LER	fLER 1973	fLER 1974[a]	fLER 1975
Grass and Cluster bean	1.97	4.43	1.00	3.63
Grass and Cowpea	1.46	3.44	1.00	2.09
Grass and Dewgram	1.62	5.14	1.00	2.34
Grass and Greengram	1.53	5.30	1.00	1.80

[a] in 1974, Bean crops failed because of drought

2. Daulay, H.S. *et al.* (1970), Rajasthan, India
Intercrops used: Guar, Mothbean, Mungbean

Results:	LER
MC Grass 2900kg/ha dry fodder	
Grass and Guar	1.10
Grass and Mothbean	1.29
Grass and Mungbean	1.20

Maize

1. Abraham, P.P. (1973), Malaysia
Intercrops used: Rubber, Soybean
Variables tested: proportion of intercrops, three-crop combination

Results:	LER
MC Maize 5542kg/ha; MC Soybean 1618kg/ha	
4 rows Maize and 6 rows Soybean	1.19
6 rows Maize and 4 rows Soybean	1.24
* Maize and Soybean in Rubber – Rubber girth stretched	

2. Ahmed, S. and H.P.M. Gunasena (1979), Nigeria; Hawaii
Intercrops used: Cowpea, Mungbean, Soybean
Variables tested: country, N fertilizer
Results:

	per cent of recommended N fertilizer dose added	LER	fLER
Nigeria:			
Maize and Cowpea:			
MC Maize 4.84tons/ha	100	1.02	1.03

MC Maize 4.35tons/ha	50	0.95	1.09
MC Maize 3.06tons/ha	0	0.97	0.95

Hawaii, USA:
Maize and Cowpea:

MC Maize 10.16tons/ha	100	1.02	1.07
MC Maize 10.06tons/ha	50	0.98	1.01
MC Maize 7.88tons/ha	25	1.06	1.14

Maize and Mungbean:

MC Maize 1.67tons/ha	100	0.87	1.62
MC Maize 1.11tons/ha	50	1.00	2.48
MC Maize 0.60tons/ha	0	1.40	5.78

Maize and Soybean:

MC Maize 6.70tons/ha	100	1.07	1.35
MC Maize 6.13tons/ha	50	1.01	1.33
MC Maize 5.52tons/ha	0	1.03	1.12

3. Akhanda, A.M. *et al.* (1978), Florida, USA
Intercrops used: Peanut, Sweet potato
Variables tested: density, Peanut and Sweet potato varieties
Results: LER
MC Maize early variety 6745kg/ha grain (average);
MC Maize medium variety 8070kg/ha;
MC Maize late maturity 8890kg/ha
The following crops were intercropped as relay crops with maize. Maize yields were not affected; thus the extra yields are the yields of the second crops.

Peanut 330kg/ha–91cm spacing between rows	1.17
Peanut 310kg/ha–46cm spacing between rows	1.18
Other Peanut varieties	1.23 , 1.17
	1.15 , 1.18
Sweet potato 2230kg/ha–91cm spacing between rows	1.17
Other Sweet potato varieties	1.10

4. Alexander, M.W. and C.F. Gentner (1962), Virginia, USA
Intercrops used: Soybean
Variables tested: density
Results: LER

10 000 plants/acre:
Maize 101.6bu/acre (0.59) and Soybean 533.0bu/acre (0.51)   1.10
15 000 plants/acre:
Maize 114.0bu/acre (0.60) and Soybean 532.2bu/acre (0.50)   1.10
20 000 plants/acre:
Maize 126.8bu/acre (0.64) and Soybean 532.8bu/acre (0.50)   1.14

5. Baker, E.F.I. (1974), Nigeria
Intercrops used: Cowpea, Millet
Results: LER
MC Maize 9738kg/ha
* Maize (1.17) and Cowpea and Millet (0.84)   2.01

6. Beets, W.C. (1975), Thailand
Intercrops used: Soybean
Results: LER   fLER
MC Maize 4202kg/ha, US $252
Maize and Soybean, by weight   0.93
Maize and Soybean, monetary value US $328   1.30

7. Beste, C.E. (1976), Maryland, USA
Intercrops used: Soybean
Variables tested: density
Experimental parameters: plant spacing in rows was varied as indicated for the two crops; the maize variety used was sweet corn; maize yield figures are for sweet corn in husk
Results: LER

1972
Maize 18cm, Soybean 7cm:
  Maize 126kg/ha (0.28) and Soybean 1735kg/ha (0.62)   0.90
Maize 34cm, Soybean 17cm:
  Maize 3350kg/ha (0.75) and Soybean 1561kg/ha (0.56)   1.31
Maize 32cm, Soybean 20cm:
  Maize 3530kg/ha (0.79) and Soybean 1308kg/ha (0.47)   1.26

1973
Maize 25cm, Soybean 14cm:
  Maize 12 090kg/ha (0.80) and Soybean 1600kg/ha (0.47)   1.27
Maize 25cm, Soybean 20cm:
  Maize 11 890kg/ha (0.79) and Soybean 1290kg/ha (0.38)   1.17
Maize 20cm, Soybean 20cm:
  Maize 14 090kg/ha (0.93) and Soybean 1590kg/ha (0.46)   1.39

8. Chowdhury, S.L. (1979), Ranchi, India
Intercrops used: Pigeon pea
Results: LER   fLER
Maize 2860kg/ha (0.99) and Pigeon pea 620kg/ha (0.97)   1.96   1.96
Maize 2280kg/ha (1.09) and Pigeon pea 1190kg/ha (0.44)   1.53   1.45
Maize 4050kg/ha (0.83) and Pigeon pea 1200kg/ha (0.57)   1.40   1.44

9. Chowdhury, S.L. (1979), Samba, India
Intercrops used: Cowpea
Results: LER   fLER
Maize 2370kg/ha (0.66) and Cowpea 440kg/ha (0.45)   1.11   1.22
Maize 1020kg/ha (0.77) and Cowpea 830kg/ha (0.52)   1.29   1.23

10. Chowdhury, S.L. (1979), Indore, India
Intercrops used: Peanut, Soybean

Results:	LER	fLER
Maize 3170kg/ha (0.75) and Peanut 1850kg/ha (0.29)	1.04	1.00
Maize 3170kg/ha (0.77) and Soybean 2240kg/ha (0.35)	1.12	1.05

11. Cordero, A. and R.E. McCollum (1979), N. Carolina, USA
Intercrops used: Snap bean, Soybean, Sweet potato
Variables tested: density, maize varieties, N fertilizer, two- and three-crop combinations
Experimental parameters: normal spacing was 97cm between maize rows; double rows alternated between 46cm and 147cm between maize rows; intercrops were planted in the 147cm space

Results:	LER
No N added	
Normal Spacing:	
Maize 4.01tons/ha (1.03) and Soybean 1.14tons/ha (0.41)	1.44
Double rows of Maize:	
Maize 4.18tons/ha (1.09) and Soybean 1.64tons/ha (0.59)	1.68
Normal spacing:	
Maize 2.86tons/ha (1.01) and Sweet potato 2.28tons/ha (0.08)	1.09
Double rows of Maize:	
Maize 4.46tons/ha (1.23) and Sweet potato 1.54tons/ha (0.05)	1.28
Double rows tall Maize:	
Maize 3.60tons/ha (0.63) and Snap bean 3.41tons/ha (0.72)	1.35
Double rows short Maize:	
Maize 2.74tons/ha (0.53) and Snap bean 2.90tons/ha (0.61)	1.14
* *Three crops*: Maize and Spring bean and Fall bean	1.73
Double rows tall Maize:	
Maize 4.60tons/ha (0.81) and Soybean 1.21tons/ha (0.42)	1.23
Double rows short Maize:	
Maize 4.86tons/ha (0.93) and Soybean 1.04tons/ha (0.36)	1.29
168kg/ha N added	
Normal spacing:	
Maize 7.18tons/ha (0.94) and Soybean 0.64tons/ha (0.22)	1.16
Double rows of Maize:	
Maize 5.50tons/ha (0.84) and Soybean 1.70tons/ha (0.58)	1.42
180kg/ha N added	
Normal spacing:	
Maize 8.60tons/ha (1.00) and Sweet potato 4.17tons/ha (0.14)	1.14
Double rows of Maize:	
Maize 8.10tons/ha (0.91) and Sweet potato 4.30tons/ha (0.15)	1.06

12. Crookston, R.K. (1976), Minnesota, USA
Intercrops used: Soybean
Experimental parameters: crop replacement was the intercropping method used; results are quoted from various locations

Results:	LER
Pairs of rows: Maize and Soybean	1.20

Appendix I: Land equivalent ratios

Quotes	LER	
Borst and Park – Ohio, USA: Maize and Soybean	1.2	
Beste – Maryland, USA: Maize and Soybean	0.9	to 1.4
Cunard – Pennsylvania, USA: Maize and Soybean	1.2	to 1.5
Philippines: Maize and Soybean	1.4	to 1.6
Indonesia: Maize and Soybean	1.2	to 1.3
India: Maize and Soybean	1.2	to 1.4
Maize and Soybean planted together and harvested together – too much competition, short growing season	0.99	to 1.0

13. Crookston, R.K. (1976), Minnesota, USA
Intercrops used: Dry bean, Peanut, Rice, Soybean, Sugarbeet

Results:	LER
Canada: Maize and Sugarbeet	1.4
Colombia: Maize and Dry bean	1.4
India: Maize and Soybean	1.4
Minnesota, USA: Maize and Soybean	1.2
Philippines: Maize and Rice	1.6
Taiwan: Maize and Peanut	1.5

14. Dalal, R.C. (1974), Trinidad
Intercrops used: Pigeon pea
Variables tested: planting pattern

Results:	LER
MC Maize 3230kg/ha; MC Pigeon pea 1870kg/ha	
Mixed on each hill: Maize (0.54) and Pigeon pea (1.08)	1.62
Alternate rows: Maize (0.83) and Pigeon pea (0.98)	1.81

15. Dalal, R.C. (1977), Trinidad
Intercrops used: Soybean
Variables tested: N fertilizer, planting pattern
Results: LER

MC Maize: 5082kg/ha with no N added; 5623kg/ha with 100kg/ha N added, an 11 per cent increase in yield
MC Soybean: 1478kg/ha with no N added; 1789kg/ha with 100kg/ha N added, a 21 per cent increase in yield

No N added:	
Mixed in same row: Maize (0.85) and Soybean (0.11)	0.96
Alternate rows: Maize (0.96) and Soybean (0.79)	1.75
2 rows Maize alternating with 2 rows Soybean: Maize (0.95) and Soybean (0.21)	1.16

100kg/ha N added:	
Mixed in same row: Maize (0.83) and Soybean (0.10)	0.93
Alternate rows: Maize (0.95) and Soybean (0.28)	1.23
2 rows maize alternating with 2 rows soybean: Maize (0.94) and Soybean (0.17)	1.11

16. Finlay, R.C. (1974), Tanzania
Intercrops used: Cowpea, Soybean
Variables tested: planting pattern

Results:	LER
Same row: Maize and Cowpea	1.00
Same row: Maize and Soybean	1.05
Alternate rows: Maize and Soybean	0.92

17. Francis, C.A. *et al.* (1982a), Colombia
Intercrops used: Bean
Variables tested: bean varieties, timing
Experimental parameters: Maize and 4 varieties of bean were used; Maize was always planted the same date, Bean at 5 different dates

Results:	LER

Bean type I, determinate Bush bean
First season
MC Maize 6380kg/ha; MC Bean 1786kg/ha

Bean planted 10 days earlier: Maize (0.41) and Bean (0.91)	1.32
Bean planted 5 days earlier: Maize (0.95) and Bean (0.52)	1.47
Simultaneous planting: Maize (1.16) and Bean (0.36)	1.52
Bean planted 5 days later: Maize (1.10) and Bean (0.39)	1.49
Bean planted 10 days later: Maize (1.10) and Bean (0.34)	1.44

Second season
MC Maize 6530kg/ha; MC Bean 1512kg/ha

Bean planted 10 days earlier: Maize (0.69) and Bean (0.75)	1.44
Bean planted 5 days earlier: Maize (0.85) and Bean (0.49)	1.34
Simultaneous planting: Maize (0.82) and Bean (0.54)	1.36
Bean planted 5 days later: Maize (0.96) and Bean (0.40)	1.36
Bean planted 10 days later: Maize (1.03) and Bean (0.35)	1.38

Bean type II, semi-indeterminate Bush bean
First season
MC Maize 6380kg/ha; MC Bean 2420kg/ha

Bean planted 10 days earlier: Maize (0.17) and Bean (0.88)	1.05
Bean planted 5 days earlier: Maize (0.72) and Bean (0.67)	1.39
Simultaneous planting: Maize (1.01) and Bean (0.46)	1.47
Bean planted 5 days later: Maize (0.86) and Bean (0.53)	1.39
Bean planted 10 days later: Maize (0.91) and Bean (0.42)	1.33

Second season
MC Maize 6530kg/ha; MC Bean 1985kg/ha

Bean planted 10 days earlier: Maize (0.67) and Bean (0.81)	1.48
Bean planted 5 days earlier: Maize and Bean	1.48
Simultaneous planting: Maize (0.84) and Bean (0.51)	1.35
Bean planted 5 days later: Maize (0.99) and Bean (0.41)	1.40
Bean planted 10 days later: Maize (1.03) and Bean (0.33)	1.36

Bean type III, indeterminate, non-climbing Bean
First season
MC Maize 6380kg/ha; MC Bean 2005kg/ha

Bean planted 10 days earlier: Maize (0.32) and Bean (0.81)	1.13
Bean planted 5 days earlier: Maize (0.79) and Bean (0.55)	1.34

Simultaneous planting: Maize (0.91) and Bean (0.44)   1.35
Bean planted 5 days later: Maize (0.92) and Bean (0.37)   1.29
Bean planted 10 days later: Maize (0.88) and Bean (0.29)   1.17

Second season
MC Maize 6530kg/ha; MC Bean 1915kg/ha
Bean planted 10 days earlier: Maize (0.62) and Bean (0.78)   1.40
Bean planted 5 days earlier: Maize and Bean   1.43
Simultaneous planting: Maize (0.78) and Bean (0.55)   1.33
Bean planted 5 days later: Maize (0.89) and Bean (0.42)   1.31
Bean planted 10 days later: Maize (0.97) and Bean (0.33)   1.30

Bean type IV, indeterminate, climbing bean
First season
MC Maize 6380kg/ha; MC Bean 2541kg/ha
Bean planted 10 days earlier: Maize (0.09) and Bean (0.43)   0.52
Bean planted 5 days earlier: Maize (0.47) and Bean (0.56)   1.03
Simultaneous planting: Maize (0.69) and Bean (0.53)   1.22
Bean planted 5 days later: Maize (0.80) and Bean (0.36)   1.16
Bean planted 10 days later: Maize (0.86) and Bean (0.41)   1.27

Second season
MC Maize 6530kg/ha; MC Bean 3949kg/ha
Bean planted 10 days earlier: Maize (0.45) and Bean (0.43)   0.88
Bean planted 5 days earlier: Maize (0.60) and Bean (0.46)   1.06
Simultaneous planting: Maize (0.76) and Bean (0.41)   1.17
Bean planted 5 days later: Maize (0.87) and Bean (0.30)   1.17
Bean planted 10 days later: Maize (0.99) and Bean (0.31)   1.30

18. Francis, C.A. *et al.* (1982b), Colombia
Intercrops used: Bean
Variables tested: density, maize varieties
Results:

Tall maize

Bean density (plants/m^2)	LERs Maize density (plants/m^2)			
	0.95	2.70	5.61	9.75
15.2	1.32	1.50	1.04	0.89
17.8	1.14	1.65	1.07	0.85
24.3	1.21	1.57	1.09	0.74
40.2	1.18	1.60	1.09	0.97

Dwarf maize

Bean density (plants/m^2)	LERs Maize density (plants/m^2)			
	2.01	3.48	5.96	7.08
10.0	1.14	1.17	1.10	0.98
19.0	1.20	1.21	1.18	1.07
29.2	1.18	1.24	1.07	1.28
35.2	1.18	1.25	1.15	1.27

19. Gangwar, B. and G.S. Kalkar (1982), Uttar Pradesh, India
Intercrops used: Blackgram, Cowpea, Greengram, Peanut, Soybean
Results:

	LER	fLER
MC Maize 1410kg/ha (av. of 2 yrs) Net return/rupee = 0.51	1.00	1.00
Maize (1.18) and Blackgram	1.63	2.58
Maize (1.18) and Cowpea	1.52	2.54
Maize (1.27) and Greengram	1.58	2.44
Maize (1.15) and Peanut	1.91	2.78
Maize (1.04) and Soybean	1.30	1.68

20. Herrara, W.A. *et al.* (1979), Philippines
Intercrops used: Peanut, Rice
Variables tested: density, proportion of intercrops
Experimental parameters: all peanut rows were spaced 50cm apart; all rice rows were spaced 25cm apart
Results:

Maize and Peanut
MC Maize 4.25tons/ha; MC Peanut 1.60tons/ha

		Rows	#plants/ha				#plants/ha	LER
4 rows	Maize	1m	20 000	and 8 rows	Peanut		80 000	1.24
3 rows	Maize	1m	20 000	and 6 rows	Peanut		80 000	1.38
2 rows	Maize	1m	20 000	and 4 rows	Peanut		80 000	1.28
1 row	Maize	1m	20 000	and 2 rows	Peanut		80 000	1.38
1 row	Maize	50cm	20 000	and 1 row	Peanut		80 000	1.30
1 row	Maize	1m	8 000	and 8 rows	Peanut		128 000	1.32
3 rows	Maize	1m	17 142	and 8 rows	Peanut		91 428	1.45
4 rows	Maize	1m	22 856	and 6 rows	Peanut		68 571	1.24
4 rows	Maize	1m	32 000	and 2 rows	Peanut		32 000	1.26
solid	Maize	2m	30 000	and solid	Peanut		160 000 (N–S)	1.50
solid	Maize	2m	30 000	and solid	Peanut		160 000 (E–W)	1.49
solid	Maize	1m	30 000	and solid	Peanut		160 000	1.53
solid	Maize	1m	60 000	and solid	Peanut		160 000	1.68

Maize and Rice
MC Maize 17.3tons/ha; MC Rice 18.5tons/ha

		Rows	#plants/ha				Seed rate	LER
8 rows	Maize	50cm	20 000	and 16 rows	Rice		50kg/ha	0.93
6 rows	Maize	50cm	20 000	and 12 rows	Rice		50kg/ha	1.09
4 rows	Maize	50cm	20 000	and 8 rows	Rice		50kg/ha	1.07
2 rows	Maize	50cm	20 000	and 4 rows	Rice		50kg/ha	1.04
1 row	Maize	50cm	20 000	and 2 rows	Rice		50kg/ha	1.25
2 rows	Maize	50cm	8 000	and 16 rows	Rice		80kg/ha	1.00
6 rows	Maize	50cm	17 142	and 16 rows	Rice		57kg/ha	0.98
8 rows	Maize	50cm	22 856	and 12 rows	Rice		43kg/ha	0.97
8 rows	Maize	50cm	32 000	and 4 rows	Rice		20kg/ha	1.00
1 row	Maize	2m	30 000	and 7 rows	Rice		88kg/ha	1.43
1 row	Maize	2m	30 000	and 7 rows	Rice		88kg/ha	1.49
1 row	Maize	1m	30 000	and 3 rows	Rice		75kg/ha	0.95
1 row	Maize	1m	60 000	and 3 rows	Rice		75kg/ha	0.99

21. IRAT #359 (1975), Togo
Intercrops used: Bean
Results: LER
MC Maize 2398kg/ha; MC Bean 937kg/ha
Maize (0.97) and Bean (0.27)   1.24
2ha Maize and Bean monocropped yielded 3339kg/ha
2ha Maize and Bean (intercropped) yielded 5186kg/ha   1.55

22. Jain, T.C. and G.N. Rao (1980), Andhra Pradesh, India
Intercrops used: Cowpea, Greengram, Peanut
Variables tested: density
Experimental parameters: data supplied by Reddy
Results:
LERs are given for *3 different spacings of Maize*:

	60cm × 30cm	75cm × 24cm	90cm × 20cm
Maize and Cowpea	1.33	1.33	1.37
Maize and Greengram	1.38	1.40	1.43
Maize and Peanut	1.33	1.39	1.41

23. Khehra, A.S. *et al.* (1979), Punjab, India
Intercrops used: Blackgram
Results: LER   fLER
Maize and Blackgram – single rows   1.02   1.11

24. King, C. *et al.* (1978), Florida, USA
Intercrops used: Soybean
Variables tested: proportion of intercrops
Experimental parameters: crop replacement intercropping was used; crops were irrigated
Results: LER
MC Maize 179bu/acre; MC Soybean 26.4bu/acre
2 rows Maize (0.87) and 4 rows Soybean (0.35)   1.22
MC Maize 157bu/acre; MC Soybean 27bu/acre
4 rows Maize (0.77) and 4 rows Soybean (0.36)   1.13

25. King, C. *et al.* (1978), Florida, USA
Intercrops used: Soybean
Variables tested: proportion of intercrops
Experimental parameters: crops were not irrigated; rows were 20 inches apart; crop replacement intercropping was used, i.e. rows of maize were replaced by rows of soybean; results are an average of 3 years
Results: LER
MC Maize 105bu/acre; MC Soybean 47bu/acre
2 rows Maize 172bu/acre and 4 rows Soybean 34bu/acre   1.41
4 rows Maize 141bu/acre and 2 rows Soybean 34bu/acre   1.16
4 rows Maize and 4 rows left empty   0.77

26. Krantz, B.A. (1979), Hyderabad, India
Intercrops used: Pigeon pea
Results: fLER
MC Maize Rs2730/ha
Maize and Pigeon pea – Rs4920/ha   1.57
Maize and Pigeon pea – Rs5040/ha   1.75

Maize and Pigeon pea – Rs4210/ha	1.83
Maize and Pigeon pea – Rs4260/ha	1.53
Maize and Pigeon pea – Rs3590/ha	1.82
Maize and Pigeon pea – Rs4090/ha	1.70
Maize and Pigeon pea – Rs3450/ha	2.20
Maize and Pigeon pea – Rs4190/ha	1.83

27. May, K.W. and R. Misangu (1980), Tanzania
Intercrops used: Cowpea, Legume soybean
Variables tested: planting pattern
Experimental parameters: maize and the intercrop were planted in the same hole and in alternate holes
Results:

	LERs	
	Alternate holes	Same hole
Maize and Cowpea	1.27	1.37
Maize and Legume	1.26	1.34
Maize and Soybean	1.27	1.32

28. Mohta, N.K. and R. De (1980), New Delhi, India
Intercrops used: Soybean
Variables tested: density, planting pattern, proportion of intercrops
Results:

	LER
MC Maize 3.34tons/ha – rows 60cm apart	
MC Maize 2.41tons/ha – rows 120cm apart	
Alternate rows, 30cm apart: Maize and Soybean	1.13
Alternate pairs of rows, all rows 30cm apart: Maize and Soybean	1.24
Alternate rows, 60cm apart: Maize and Soybean	1.42
1 row Maize @ 120cm and 2 rows Soybean	1.48
1 row Maize @ 120cm and 3 rows Soybean	1.43

29. Nadar, H.M. and G.E. Rodewald (1979), Kenya
Intercrops used: Bean
Variables tested: density, soil preparation, timing
Experimental parameters: MC maize was sown 1 plant per hole; maize and bean were planted 1 maize and 1 bean plant per hole; ploughing after maize harvest and then planting bean (till) was compared with relay planting (no till); relay planting enables beans to take better advantage of rain
Results:

Row Spacing	Soil Preparation	MC Maize	Maize and Soybean	LER
60cm	Till	4748kg/ha	5403kg/ha	1.14
60cm	No Till	4663kg/ha	5712kg/ha	1.24
75cm	Till	5243kg/ha	5240kg/ha	1.00
75cm	No Till	4347kg/ha	5833kg/ha	1.16
90cm	Till	3946kg/ha	4753kg/ha	1.20
90cm	No Till	3423kg/ha	5281kg/ha	1.54

30. Oelsligle, D.D. *et al.* (1976), Costa Rica
Intercrops used: Cassava
Experimental parameters: no N fertilizer was added
Results: LER
MC Maize 1.93tons/ha; MC Cassava 24.3tons/ha
Maize (1.42) and Cassava (1.18)   2.60

31. Pendleton, J.W. *et al.* (1963), Illinois, USA
Intercrops used: Soybean
Results: LER
2 rows Maize (0.30) and 2 rows Soybean (1.00)   1.30

32. Raynolds, M.K. and R.W. Elias (1980), Virginia, USA
Intercrops used: Soybean
Variables tested: density
Results: LER
Higher Maize yields resulted in lower Soybean yields   1.00

33. Reddy, K.A. *et al.* (1980), Hyderabad, India
Intercrops used: Blackgram, Cowpea, Finger millet, Foxtail millet, Greengram, Mustard, Peanut, Safflower, Sesame, Soybean
Variables tested: planting pattern
Experimental parameters: Maize and intercrops were planted in paired rows
Results:

	LER	fLER
MC Maize 41.9q/acre, *winter 1975–76*		
MC Maize planted in paired rows gave 8–12 per cent increase in yield		
Maize and Mustard	1.18	1.76
Maize and Peanut	1.05	1.24
Maize and Safflower	1.11	1.31
Maize and Sesame	0.78	0.78
Maize and Soybean	0.77	0.84
Maize and Wheat	1.27	1.38
MC Maize 30.1q/ha, *1976*		
Maize and Blackgram	1.01	1.03
Maize and Cowpea	1.02	1.04
Maize and Finger millet	1.03	1.06
Maize and Foxtail millet	1.06	1.08
Maize and Greengram	1.02	1.04
Maize and Peanut	1.32	1.56
Maize and Safflower	0.97	1.18
Maize and Soybean	1.05	1.07

34. Reddy, G.J. and M.R. Reddi (1981), Hyderabad, India
Intercrops used: Cowpea, Greengram, Peanut
Variables tested: density
Experimental parameters: maize density was varied; three densities of maize were tested with three different intercrops
Results: LER   fLER
MC Maize 60cm × 30cm – 4120kg/ha
MC Maize 75cm × 24cm – 4273kg/ha
MC Maize 90cm × 20cm – 4080kg/ha
MC Cowpea 1074kg/ha; MC Greengram 878kg/ha; MC Peanut 2072kg/ha

Maize 60cm × 30cm (0.90) and Cowpea (0.43)	1.33	1.15
Maize 60cm × 30cm (0.97) and Greengram (0.40)	1.37	1.24
Maize 60cm × 30cm (0.94) and Peanut (0.39)	1.33	1.31
Maize 75cm × 24cm (0.88) and Cowpea (0.45)	1.33	1.13
Maize 75cm × 24cm (0.98) and Greengram (0.42)	1.40	1.24
Maize 75cm × 24cm (0.94) and Peanut (0.45)	1.39	1.35
Maize 90cm × 20cm (0.91) and Cowpea (0.46)	1.37	1.18
Maize 90cm × 20cm (0.98) and Greengram (0.45)	1.43	1.26
Maize 90cm × 20cm (0.94) and Peanut (0.47)	1.41	1.40

35. Remison, S.U. (1980), Nigeria
Intercrops used: Cowpea
Variables tested: Cowpea cultivars, proportion of intercrops
Results: LER
1977
MC Maize 6.25tons/ha; MC Cowpea (Ife Brown cultivar)
    1.83 tons/ha; MC Cowpea (New Era cultivar) 0.67 tons/ha
Ife Brown Cowpea cultivar
67 per cent Maize and 33 per cent Cowpea   0.87
50 per cent Maize and 50 per cent Cowpea   0.92
33 per cent Maize and 67 per cent Cowpea   0.94

New Era Cowpea cultivar
67 per cent Maize and 33 per cent Cowpea   1.72
50 per cent Maize and 50 per cent Cowpea   2.35
33 per cent Maize and 67 per cent Cowpea   1.56

1978
MC Maize 2.91tons/ha; MC Cowpea (Ife Brown cultivar)
    0.67 tons/ha; MC Cowpea (TVU cultivar) 0.58 tons/ha
Ife Brown Cowpea cultivar
67 per cent Maize and 33 per cent Cowpea   0.87
50 per cent Maize and 50 per cent Cowpea   0.99
33 per cent Maize and 67 per cent Cowpea   0.99

TVU 4557 Cowpea cultivar
67 per cent Maize and 33 per cent Cowpea   1.00
50 per cent Maize and 50 per cent Cowpea   1.08
33 per cent Maize and 67 per cent Cowpea   1.22

36. Searle, P.G. *et al.* (1981), Australia
Intercrops used: Peanut, Soybean
Variables tested: N fertilizer
Results:

kg/ha N added	Maize and Peanut – LER	Maize and Soybean – LER
0	1.37	1.36
25	1.12	1.03
50	1.22	1.01
100	1.29	1.09

Appendix I: Land equivalent ratios

37. Searle, P.G. *et al.* (1981), Australia
Intercrops used: Millet, Peanut, Soybean
Variables tested: N fertilizer
Results:
MC Millet 7700kg/ha; MC Peanut 1741kg/ha; MC Soybean 2677kg/ha

LER (yield)

N added	Maize and Millet	Maize and Peanut	Maize and Soybean
0kg/ha	1.00 (7700kg/ha)	1.37 (530kg/ha)	1.36 (534kg/ha)
25kg/ha	1.00 (7700kg/ha)	1.12 (447kg/ha)	1.03 (563kg/ha)
50kg/ha	1.00 (7700kg/ha)	1.22 (418kg/ha)	1.01 (569kg/ha)
100kg/ha	1.00 (7700kg/ha)	1.29 (378kg/ha)	1.09 (532kg/ha)

38. Sharma, K.N. *et al.* (1979), Punjab, India
Intercrops used: Blackgram, Greengram
Variables tested: fertilizer
Experimental parameters: 100 per cent of the recommended fertilizer dose was used for maize; the dose was varied for blackgram and greengram
Results: *LER*
MC Maize 2 431kg/ha

	LER
Maize (100 per cent dose) and Blackgram (no fertilizer)	0.89
Maize (100 per cent dose) and Blackgram (25 per cent dose)	1.24
Maize (100 per cent dose) and Blackgram (50 per cent dose)	1.11
Maize (100 per cent dose) and Greengram (no fertilizer)	1.04
Maize (100 per cent dose) and Greengram (25 per cent dose)	1.12
Maize (100 per cent dose) and Greengram (50 per cent dose)	1.03

39. Singh, C.M. and W.S. Guleria (1979), Himachal Pradesh, India
Intercrops used: Soybean
Variables tested: N fertilizer
Results: *LER*
MC Maize 1487kg/ha
IC Maize 1608kg/ha

	LER
Maize and Soybean – no N added	1.91
Maize and Soybean – N added	1.82

40. Singh, C.M. and P. Chand (1980), Himachal Pradesh, India
Intercrops used: Soybean
Variables tested: N fertilizer
Results: *LER*   *fLER*
MC Maize 23.5q/ha

	LER	fLER
Maize and Soybean (average)	1.16	2.00

N added		Net profit	
0kg/ha	Maize and Soybean	Rs861/ha	
40kg/ha	Maize and Soybean	Rs705/ha	13.35
60kg/ha	Maize and Soybean	Rs1323/ha	3.02
80kg/ha	Maize and Soybean	Rs1371/ha	1.48
120kg/ha	Maize and Soybean	Rs1271/ha	1.39

Note that benefit from intercropping with soybean decreased as more N was added.

41. Wahua, T.A.T. et al. (1981), Nigeria
Intercrops used: Cowpea
Variables tested: Maize cultivars
Experimental parameters: 3 cultivars of maize were tested:
KEWE – short, early maturing, erect, small leaves: NCD tall, late maturing, high yielding, large floppy leaves, and large ABCD – short, later maturing, large erect leaves

Results: LER

Early planting
KEWE Maize 4518kg/ha (0.93) and Cowpea 2833kg/ha (0.46)   1.39
NCD Maize 6575kg/ha (1.02) and Cowpea 2833kg/ha (0.37)   1.39
large ABCD Maize 5356kg/ha (0.94) and Cowpea 2833kg/ha (0.60)   1.54

Late season
KEWE Maize 3441kg/ha (0.97) and Cowpea 1323kg/ha (0.50)   1.47
NCD Maize 5370kg/ha (1.01) and Cowpea 1323kg/ha (0.48)   1.49
large ABCD Maize 5071kg/ha (0.98) and Cowpea 1323kg/ha (0.40)   1.38

42. Wijesinha, A. et al. (1982), Brazil
Intercrops used: Bean
Variables tested: Bean and Maize cultivars
Experimental parameters: there were 4 bean cultivars: A, B, C, and D; and 2 maize cultivars: X and Y.
Results:

	LERs			
	A	B	C	D
X	1.40	1.56	1.41	1.50
Y	1.17	1.74	1.27	1.05

Oat

1. Baeumer, K. and C.T. DeWit (1968), Holland
Intercrops used: Barley, Long pea, Short pea
Experimental parameters: Pea yields are measured as weight of dry matter
Results: LER
MC Oat 78.9kg/100m^2 @ 25cm spacing;
MC Long pea 60.5kg/100m^2; MC Short pea 64.7kg/100m^2
Oat and Long pea   1.10
Oat and Short pea   1.21

1960: MC Oat 83.0kg/100m^2; MC Barley 93.6kg/100m^2
Oat and Barley   1.11

1961: MC Oat 88.9kg/100m^2 @ 25cm; MC Barley 82.3kg/100m^2
Oat and Barley   1.00

1963: MC Oat 96.0kg/100m^2; MC Short pea 75.6kg/100m^2
Oat and Short pea   1.02

1964: MC Oat 102.4kg/100m^2; MC Short pea 94.6kg/100m^2
Oat and Short pea   1.00

2. Bengtsson, A. (1973), Sweden
Intercrops used: Pea
Variables tested: N fertilizer, proportion of intercrops
Results: LER
No N added – 20 per cent Oat in Pea mixture is best
MC Oat 2450kg/ha; MC Pea 2780kg/ha
Oat (0.44) and Pea (0.73)   1.17

30kg/ha N added – 40 per cent Oat in Pea mixture is best
MC Oat 2940kg/ha; MC Pea 2810kg/ha
Oat (0.65) and Pea (0.49)   1.14

60kg/ha N added – 60 per cent Oat in Pea mixture is best
MC Oat 3330kg/ha; MC Pea 2900kg/ha
Oat (0.82) and Pea (0.26)   1.08

3. Brown, C.M. and D.W. Graffis (1976), Illinois, USA
Intercrops used: Soybean
Results: LER
MC Oat silage 4065kg/ha; MC Soybean 136bu/acre
Oat silage (1.05) and Soybean (0.72)   1.77
MC Oat grain 234bu/ha; MC Soybean 69bu/ha
Oat grain (0.87) and Soybean (0.51)   1.38

4. Chan, L.M. and C.M. Brown (1980), Illinois, USA
Intercrops used: Soybean
Variables tested: density
Results: 1977   LER

41cm spacing: MC Oat 3218kg/ha: MC Soybean 3593kg/ha
Oat (0.99) and Soybean (0.51)   1.50

61cm spacing: MC Oat 2117kg/ha: MC Soybean 3443kg/ha
Oat (1.05) and Soybean (0.72)   1.77

81cm spacing: MC Oat 1859kg/ha; MC Soybean 3651kg/ha
Oat (1.08) and Soybean (0.74)   1.82

5. Chan, L.M. and C.M. Brown (1980), Illinois, USA
Intercrops used: Soybean
Variables tested: density
Experimental parameters: relay cropping was used
Results: 1976   LER
21 cm spacing: MC Oat 4545kg/ha; MC Soybean 3414kg/ha
Oat (0.95) and Soybean (0.47)   1.42

41cm spacing: MC Oat 4365kg/ha; MC Soybean 2876kg/ha
Oat (0.99) and Soybean (0.71)   1.70

61cm spacing: MC Oat 3338kg/ha; MC Soybean 3266kg/ha
Oat (0.91) and Soybean (0.63)   1.54

81cm spacing: MC Oat 2702kg/ha; MC Soybean 3158kg/ha
Oat (1.01) and Soybean (0.80)   1.81

Peanut

1. Appadurai, R. and R.K.V. Selva (1974), Madras, India
Intercrops used: Pigeon pea
Results: *f*LER
1st season (average of 3 years): Peanut and Pigeon pea   1.98
2nd season (average of 3 years): Peanut and Pigeon pea   1.09

2. Baker, E.F.I. (1974), Nigeria
Intercrops used: Cowpea, Millet, Sorghum
Variables tested: four-crop combination
Results: LER
MC Peanut 587kg/ha
* Peanut and Cowpea and Millet and Sorghum   2.23

3. Baker, E.F.I. (1978), Nigeria
Intercrops used: Maize, Millet, Sorghum
Variables tested: two-, three- and four-crop combinations

Results:	LER	*f*LER
MC Peanut 2441kg/ha value Naira 352		
Peanut and Maize	1.27	1.18
Peanut and Millet	1.42	1.52
Peanut and Sorghum	1.36	1.43
* Peanut and Maize and Sorghum	1.25	1.28
* Peanut and Millet and Sorghum	1.33	1.41
* Peanut and Sorghum and Millet	1.36	1.39
* Peanut and Maize and Millet and Sorghum	1.54	1.53

4. Balasubrahmanyan, R. (1950), Andhra Padesh, India
Intercrops used: Cotton
Variables tested: Cotton cultivars, density
Results: *f*LER
Peanut and Cotton
Average of 5 experiments with various Cotton cultivars   1.73
Average of 4 experiments with various Cotton cultivars   1.09
Average of 4 experiments with various Cotton cultivars and spacings   1.44

5. Norman, D.W. (1974), Nigeria
Intercrops used: Cowpea, Millet, Sorghum
Variables tested: four-crop combination
Results: *f*LER
MC Peanut 433kg/ha
* Peanut and Cowpea and Millet and Sorghum   2.17

Pigeon pea

1. Gahlot, K.L. *et al.* (1978), Uttar Pradesh, India
Intercrops used: Greengram
Variables tested: density
Experimental parameters: results are given for 5 different densities of planting

Results:	LER	fLER
MC Pigeon pea 2086kg/ha		
Pigeon pea and Greengram	1.32	1.32
Pigeon pea and Greengram	1.22	1.21
Pigeon pea and Greengram	1.21	1.22
Pigeon pea and Greengram	1.23	1.23
Pigeon pea and Greengram	1.16	1.16

2. Hegde, D.M. and C.S. Saraf (1979), New Delhi, India
Intercrops used: Blackgram, Cowpea, Mung bean
Experimental parameters: results are an average of 2 years

Results:	LER	fLER
MC Pigeon pea 1608kg/ha		
Pigeon pea and Blackgram	1.17	1.33
Pigeon pea and Cowpea	1.27	1.38
Pigeon pea and Mung bean	1.23	1.34

3. Kaul, J.N. and H.S. Sikhon (1974), Punjab, India
Intercrops used: Blackgram, Maize, Mung bean, Peanut, Soybean

Results:	LER	fLER
MC Pigeon pea 1718kg/ha		
Pigeon pea and Blackgram	1.25	1.29
Pigeon pea and Maize	2.27	
Pigeon pea and Mung bean	1.21	1.21
Pigeon pea and Peanut	1.36	1.25
Pigeon pea and Soybean	1.09	0.96

4. Mead, R. and R.W. Willey (1980), Andhra Pradesh, India
Intercrops used: Sorghum
Variables tested: Pigeon pea genotype
Results:

Pigeon pea Genotype	Yield (kg/ha) Sorghum	Pigeon pea	LER using same Pigeon pea	LER using best Pigeon pea
1	3804	850	1.47	1.46
2	3931	842	1.56	1.49
3	3640	740	1.44	1.36
4	3630	815	1.50	1.40
5	3386	757	1.43	1.31
6	3344	885	1.48	1.37
7	3899	799	1.62	1.46
8	3381	619	1.45	1.22
9	3973	585	1.44	1.35
10	3757	619	1.45	1.31
11	3232	512	1.24	1.12
12	3500	463	1.25	1.16
13	3323	503	1.27	1.14

14	3930	661	1.58	1.38
15	3198	718	1.47	1.23
16	3645	530	1.42	1.23
17	3677	720	1.66	1.35
			Average = 1.45	Average = 1.31

5. Rao, M.P. and R.W. Willey (1980b), Andhra Pradesh, India
Intercrops used: Castor bean, Cowpea, Foxtail millet, Maize, Peanut, Pearl millet, Sorghum, Soybean
Variables tested: maturity date, soil type
Experimental parameters: crop combinations were tested on Alfisols, which do not hold much water; and on Vertisols, which have a greater capacity for holding water; in 1976 the Cowpea genotype was changed
Results: the data below are listed in order of increasingly later date of maturity of the intercrops

Alfisols	LER
1975	
Pigeon pea (0.80) and Foxtail millet (1.0)	1.80
Pigeon pea (0.45) and Pearl millet (0.85)	1.30
Pigeon pea (0.45) and Sorghum (0.75)	1.20
Pigeon pea (0.45) and Cowpea (0.65)	1.10
Pigeon pea (0.80) and Soybean (0.75)	1.55
1976	
Pigeon pea (1.0) and Foxtail millet (0.70)	1.70
Pigeon pea (0.40) and Pearl millet (1.0)	1.40
Pigeon pea (0.40) and Maize (0.90)	1.30
Pigeon pea (0.75) and Cowpea (0.80)	1.55
Pigeon pea (0.95) and Peanut (0.60)	1.55
Pigeon pea (0.55) and Castor bean (0.65)	1.20
Vertisols	
1975	
Pigeon pea (1.0) and Foxtail millet (1.0)	2.00
Pigeon pea (0.80) and Pearl millet (1.05)	1.85
Pigeon pea (0.80) and Maize (0.80)	1.60
Pigeon pea (0.60) and Cowpea (0.75)	1.35
Pigeon pea (0.85) and Peanut (0.40)	1.25
1976	
Pigeon pea (0.80) and Foxtail millet (0.90)	1.70
Pigeon pea (0.80) and Pearl millet (0.45)	1.25
Pigeon pea (0.75) and Maize (0.90)	1.65
Pigeon pea (0.85) and Cowpea (0.80)	1.65
Pigeon pea (0.90) and Peanut (0.40)	1.30

Castor bean and Pigeon pea, which have similar growth habits, did poorly together. Peanut did poorly on Vertisols. Upright Soybean was less competitive with Pigeon pea than spreading Cowpea was.

Appendix I: Land equivalent ratios

6. Rao, M.P. and R.W. Willey (1980b), various locations
Intercrops used: Sorghum
Experimental parameters: results of 40 experiments were averaged.

Results:	LER
Pigeon pea and Sorghum	1.46

7. Reddy, R.P. and P.P. Tarhalkar (1977), Andhra Pradesh, India
Intercrops used: Castor bean, Peanut, Pigeon pea, Sorghum
Variables tested: density, planting pattern, proportion of intercrops

Results:	fLER
3: 1 ratio – Pigeon pea and Castor bean	1.59
Pigeon pea and paired rows of Peanut	3.11
3: 1 ratio – Pigeon pea and Peanut	2.51
Pigeon pea and paired rows of Sorghum	1.33
Pigeon pea and wide rows of Sorghum	1.27
3: 1 ratio – Pigeon pea and Sorghum	1.24

8. Roy, R.P. *et al.* (1981), Bihar, India
Intercrops used: Blackgram, Cowpea, Finger millet, Greengram, Maize, Peanut, Soybean
Experimental parameters: results are an average of 3 years, 1974–77

Results:	LER	fLER
MC Pigeon pea 2690kg/ha		
Pigeon pea (0.97) and Blackgram	1.09	1.10
Pigeon pea (0.90) and Cowpea	0.96	0.88
Pigeon pea (0.99) and Finger millet	1.29	1.26
Pigeon pea (1.00) and Greengram	1.10	1.13
Pigeon pea (1.01) and Maize	1.59	1.53
Pigeon pea (1.05) and Peanut	1.25	1.36
Pigeon pea (1.00) and Soybean	1.15	1.09

9. Saraf, C.S. (1975), New Delhi, India
Intercrops used: Blackgram, Cowpea, Maize, Mung bean, Sorghum, Soybean
Experimental parameters: results are an average of 1972 and 1973 figures

Results:	LER	fLER
Pigeon pea and Blackgram	1.05	1.03
Pigeon pea and Cowpea	1.20	1.31
Pigeon pea and Maize	1.57	0.96
Pigeon pea and Mung bean	1.18	1.14
Pigeon pea and Sorghum	0.57	0.02
Pigeon pea and Soybean	1.11	0.99

10. Singh, H.P. (1982), Uttar Pradesh, India
Intercrops used: Blackgram, Cowpea, Maize, Mung bean, Soybean
Results:
1973: MC Pigeon pea 2342kg/ha, Rs2487/ha
1974: MC Pigeon pea 2292kg/ha, Rs4027/ha

	LER		fLER	
	1973	1974	1973	1974
Pigeon pea and Blackgram	1.27	1.23	1.43	1.09
Pigeon pea and Cowpea	1.22	0.97	1.27	0.75
Pigeon pea and Maize	1.99	2.05	1.65	1.49

Pigeon pea and Mung bean	1.20	1.02	1.34	0.88
Pigeon pea and Soybean	1.30	1.08	1.44	1.00

11. Singh, K. *et al.* (1978), Uttar Pradesh, India
Intercrops used: Bajri, Blackgram, Sunflower
Variables tested: density
Results:

	LER	fLER
50cm spacing between rows:		
MC Pigeon pea 2838kg/ha		
Pigeon pea (0.25) and Bajri	1.16	0.45
Pigeon pea (0.75) and Blackgram	1.05	1.06
Pigeon pea (0.39) and Sunflower	0.65	0.58
75cm spacing between rows:		
MC Pigeon pea 2366kg/ha		
Pigeon pea (0.45) and Bajri	1.35	0.85
Pigeon pea (0.99) and Blackgram	1.25	1.58
Pigeon pea (0.75) and Sunflower	1.03	1.30

12. Singh, K. and M. Singh (1981), Uttar Pradesh, India
Intercrops used: Bajri, Blackgram, Sunflower
Variables tested: density
Results:

	LER		fLER
	1974–75	1975–76	
1974–75: MC Pigeon pea 2744kg/ha			
1975–76: MC Pigeon pea 2933kg/ha			
50cm spacing between rows:			
Pigeon pea and Bajri	1.15	1.17	0.63
Pigeon pea and Blackgram	1.05	1.03	1.09
Pigeon pea and Sunflower	0.67	0.63	0.51
75cm spacing between rows:			
Pigeon pea and Bajri	1.36	1.33	0.92
Pigeon pea and Blackgram	1.26	1.24	1.37
Pigeon pea and Sunflower	1.05	1.02	0.99

13. Singh, M. *et al.* (1979), Uttar Pradesh, India
Intercrops used: Blackgram, Cowpea, Foxtail millet, Little millet, Mung bean, Soybean
Results:

	fLER
Pigeon pea and Blackgram	1.42
Pigeon pea and Cowpea	0.83
Pigeon pea and Foxtail millet	0.76
Pigeon pea and Little millet	0.73
Pigeon pea and Mung bean	1.18
Pigeon pea and Soybean	1.09

14. Singh, M. *et al.* (1979), Uttar Pradesh, India
Intercrops used: Blackgram, Foxtail millet, Greengram, Little millet, Soybean
Results:

	LER	fLER
MC Pigeon pea 1990kg/ha		
Pigeon pea (0.80) and Blackgram	1.23	1.64

Pigeon pea (0.61) and Foxtail millet	1.01	0.62
Pigeon pea (0.79) and Greengram	1.16	1.32
Pigeon pea (0.60) and Little millet	1.07	0.69
Pigeon pea (0.62) and Soybean	1.06	1.02

15. Singh, S. and R.C. Singh (1976), Haryana, India
Intercrops used: Bajri, Blackgram, Cowpea, Greengram, Soybean
Variables tested: Bajri and Cowpea varieties, precipitation
Results: LER   fLER

1972 – 290mm rain
MC Pigeon pea 22.05q/ha, Rs2588/ha

Pigeon pea and regular Bajri	0.36	1.29
Pigeon pea and Blackgram	1.15	1.24
Pigeon pea and regular Cowpea	0.05	1.48
Pigeon pea and Greengram	1.17	1.31
Pigeon pea and Soybean	1.11	1.08

1973 – 193mm rain
MC Pigeon pea 11.52q/ha, Rs1144/ha

Pigeon pea and short Bajri	1.82	1.37
Pigeon pea and Blackgram	1.34	1.74
Pigeon pea and short Cowpea	1.37	1.33
Pigeon pea and Greengram	1.47	1.87
Pigeon pea and Soybean	1.25	1.28

16. Veeraswamy, R. *et al.* (1974), Madras, India
Intercrops used: Peanut
Variables tested: proportion of intercrops
Experimental parameters: results are an average of 2 years of data

Results:	Compared to MC Pigeon Pea	Compared to MC Peanut	
	fLER	LER	fLER
1 row Pigeon pea and 6 rows Peanut	2.24	1.37	1.41
1 row Pigeon pea and 8 rows Peanut	2.08	1.28	1.32

17. Yadav, R.L. (1981), Uttar Pradesh, India
Intercrops used: Maize, Sugar
Variables tested: N fertilizer, relay planting
Experimental parameters: Pigeon pea and Maize were planted together, followed by a crop of sugar
Results: LER

40kg/ha N added:
* Pigeon pea 0.38tons/ha and Maize 2.70tons/ha and Sugar
  58.9tons/ha   1.28

60kg/ha N added:
* Pigeon pea 0.40tons/ha and Maize 3.16tons/ha and Sugar
  62.5tons/ha   1.37

120kg/ha N added:
* Pigeon pea 0.60tons/ha and Maize 4.00tons/ha and Sugar
  70.4tons/ha   1.49

Sugar planted following Maize and Pigeon pea yielded 43 per cent more than Sugar planted following monocropped Maize.

Potato

1. Chatterjee, B.N. *et al.* (1978), West Bengal, India
Intercrops used: Wheat
Experimental parameters: the crops were irrigated; N, P and K were added

Results:	LER	fLER
MC Potato 267q/ha (average); MC Wheat 34q/ha		
Potato (0.90) and Wheat (0.44); fLER compared to MC Potato	1.34	0.85
Potato and Wheat compared to MC Wheat		2.82

Rice

1. Chowdhury, S.L. (1979), Ranchi, India
Intercrops used: Pigeon pea

Results:	LER	fLER
Rice 3140kg/ha (0.67) and Pigeon pea 430kg/ha (0.97)	1.64	1.49
Rice 1640kg/ha (0.84) and Pigeon pea 1370kg/ha (0.80)	1.64	1.62
Rice 2290kg/ha (0.43) and Pigeon pea 680kg/ha (0.98)	1.41	1.47

2. Reddy, M.H. and B.N. Chatterjee (1973), West Bengal, India
Intercrops used: Soybean
Variables tested: precipitation
Experimental parameters: 1 row of Soybean was alternated with 1 row of Jaya rice; the rows were 20cm apart; 20kg/ha of N was added.

Results:	LER
1970-wet: MC Rice 4700kg/ha; MC Soybean 1960kg/ha	
Rice (0.02) and Soybean (0.59)	0.61[a]
1971-dry: MC Rice 2060kg/ha; MC Soybean 2280kg/ha	
Rice (1.30) and Soybean (1.00)	2.30
1971-wet: MC Rice 2960kg/ha; MC Soybean 830kg/ha	
Rice (1.14) and Soybean (0.72)	1.86
1972-dry: MC Rice 1320kg/ha; MC Soybean 3540kg/ha	
Rice (0.45) and Soybean (1.06)	1.51

[a]In 1970, a very wet year, luxuriant growth of Soybean suppressed Rice growth.

Rubber

1. Hunter, J.R. and Edilberto Camacho (1961), Costa Rica
Intercrops used:

Results:	fLER
Net return from MC Rubber US$459	1.35

2. Pillai, P.N. (1974), Kerala, India
Intercrops used: Banana, Cassava
Results:
Rubber and 400 bunches Banana/ha/yr – Rubber grew bigger faster than when monocropped
Rubber and 20–40tons Cassava/ha/yr – Rubber girth was smaller, possibly because of competition with Cassava for soil nutrients

3. Pushparajah, E. and W.P. Weng (1970), Malaysia
Intercrops used: Banana, Chilli, Maize, Peanut, Pineapple, Tapioca, Vegetable
Results: % smallholders using
                                                                  this combination

Rubber and Banana	50
Rubber and Chilli	5
Rubber and Peanut	a few
Rubber and Pineapple	10
Rubber and Tapioca	a few
Rubber and Vegetable	6

When intercropped with Rubber, 700–1800kg/ha Peanut or 1100–3400kg/ha Maize did not decrease Rubber yields.

Sesame

1. Chandrasekharan, N.R. *et al.* (1974), Madras, India
Intercrops used: Peanut, Pearl millet, Pigeon pea, Sorghum

Results:	LER	fLER
MC Sesame 178kg/ha (average of 3 years) volume Rs316/ha		
Sesame (0.46) and Peanut 690kg/ha (0.62)	1.08	1.09
Sesame (0.34) and Pearl millet 166kg/ha (0.76)	1.10	1.04
Sesame (0.99) and Pigeon pea 341kg/ha (0.37)	1.36	1.37
Sesame (0.52) and Sorghum 166kg/ha (0.64)	1.16	1.11

2. Desai, N.D. and S.N. Goyal (1980), Gujarat, India
Intercrops used: Bunch peanut, Castor bean, Soybean, Spreading peanut, Sunflower
Variables tested: density, proportion of intercrops
Experimental parameters: MC yields are an average of 3 years

Results:	LER	fLER
Rows 30cm apart: MC Sesame 345kg/ha		
Rows 40cm apart: MC Sesame 439kg/ha	1.27	1.58
Sesame (0.88) 1:1 with Bunch peanut	1.10	0.87
Sesame (0.98) 2:1 with Bunch peanut	1.17	0.89
Sesame (0.71) 1:1 with Castor bean	2.10	1.76
Sesame (0.84) 2:1 with Castor bean	1.52	1.22
Sesame (0.92) 1:1 with Soybean	1.70	1.27
Sesame (1.06) 2:1 with Soybean	1.51	1.22
Sesame (0.85) 1:1 with Spreading peanut	1.11	0.87
Sesame (0.90) 2:1 with Spreading peanut	1.10	0.79
Sesame (0.62) 1:1 with Sunflower	1.41	0.57
Sesame (0.82) 2:1 with Sunflower	1.50	0.74

Sorghum

1. Bains, S.S. (1968), India
Intercrops used: Cowpea

Results:	LER
MC Sorghum 16 830kg/ha; MC Cowpea 14 930kg/ha	
Sorghum and Cowpea	1.48

2. Baker, E.F.I. (1974), Nigeria
Intercrops used: Cowpea, Maize, Millet, Peanut, Sorghum
Variables tested: relay cropping, two- and three-crop combinations

Results:	LER	fLER
MC Sorghum 786kg/ha		
* Sorghum and Cowpea and Millet and Peanut	1.66	
MC Sorghum 2893kg/ha; MC millet 3060kg/ha;		
Sorghum (0.78) and Millet (0.75)	1.53	
MC Cowpea 998kg/ha		
* 1 row Sorghum and 1 row Millet followed by Cowpea		1.73
* 1 row Sorghum and 2 rows Millet followed by Cowpea		1.80
* 2 rows Sorghum and 2 rows Millet followed by Cowpea		1.61
MC Sorghum 5 344kg/ha		
Sorghum and Maize	1.64	
Sorghum and Maize compared to MC Maize	1.17	
Sorghum and Millet	0.92	
Sorghum and Millet	1.24	
Sorghum (0.59) and Millet (0.57)	1.16	
Sorghum and Millet compared to MC Millet	1.02	

3. Baker, E.F.I. and the IAR (1979), various locations
Intercrops used: Maize
Variables tested: proportion of intercrops
Results:

Proportion of Optimum Sorghum Sale population	LERs Proportion of optimum Maize sale population				
	0.53	0.80	1.06	1.33	1.60
0.65	0.90	0.97	1.07	1.31	1.13
0.85	1.00	1.16	1.33	1.34	1.12
1.05	1.27	1.26	1.29	1.08	1.09
1.25	1.07	1.24	1.26	1.12	1.12
1.45	1.16	1.11	1.21	1.23	1.13

Note: For MC Maize or Sorghum 1.00 is the optimum population (the population that gives the highest yield). In this table, for instance, 0.53 of the optimum Maize population plus 0.65 of the optimum Sorghum population gives a LER of 0.90.

4. Borse, R.H. *et al.* (1980), Maharashtra, India
Intercrops used: Blackgram, Cowpea, Greengram
Variables tested: Planting pattern

Results:	LER	fLER
MC Sorghum single rows 5185kg/ha		
MC Sorghum paired rows	1.03	1.05
Paired rows of Sorghum (0.91) and Blackgram	0.92	0.98
Paired rows of Sorghum (0.73) and Cowpea	0.86	0.96
Paired rows of Sorghum (1.04) and Greengram	1.04	1.07

5. Chandravanshi, B.R. (1975), Madhya Pradesh, India
Intercrops used: Mung bean, Peanut, Soybean
Results: fLER
MC Sorghum grain 3 183kg/ha
Sorghum and Mung bean   0.99
Sorghum and Peanut   0.99
Sorghum and Soybean   1.11

6. Chowdhury, S.L. (1979), Bijapur, India
Intercrops used: Safflower
Results: LER   fLER
MC Sorghum 2160kg/ha; MC Safflower 1640kg/ha
Sorghum (0.49) and Safflower (0.47)   0.96   0.95
MC Sorghum 2040kg/ha; MC Safflower 1810kg/ha
Sorghum (0.46) and Safflower (0.45)   0.91   0.90

7. Chowdhury, S.L. (1979), Kovilpatti, India
Intercrops used: Castor bean, Pigeon pea
Results: LER   fLER
MC Sorghum 1350kg/ha; MC Castor bean 170kg/ha;
MC Pigeon pea 140kg/ha
Sorghum (0.19) and Castor bean (0.82)   1.01   0.64
Sorghum (0.52) and Pigeon pea (0.48)   1.00   1.04

8. Chowdhury, S.L. (1979), Akola, India
Intercrops used: Blackgram, Greengram, Pigeon pea
Results: LER   fLER
MC Sorghum 3350kg/ha; MC Blackgram 1800kg/ha;
MC Greengram 1640kg/ha
Sorghum (0.89) and Blackgram (0.25)   1.14   1.02
Sorghum (0.88) and Greengram (0.45)   1.33   1.35
MC Sorghum 5270kg/ha; MC Pigeon pea 2670kg/ha
Sorghum (0.91) and Pigeon pea (0.23)   1.14   1.06

9. Chowdhury, S.L. (1979), Hyderabad, India
Intercrops used: Pigeon pea
Results: LER   fLER
MC Sorghum 1950kg/ha; MC Pigeon pea 1330kg/ha
Sorghum (0.62) and Pigeon pea (0.55)   1.17   1.15

10. Chowdhury, S.L. (1979), Rewa, India
Intercrops used: Pigeon pea
Results: LER   fLER
MC Sorghum 2540kg/ha; MC Pigeon pea 1030kg/ha
Sorghum (0.88) and Pigeon pea (0.46)   1.34   1.33
MC Sorghum 950kg/ha; MC Pigeon pea 680kg/ha
Sorghum (0.62) and Pigeon pea (0.35)   0.97   0.90

11. Chowdhury, M.S. and R.N. Misangu (1979), Tanzania
Intercrops used: Chickpea
Variables tested: Chickpea inoculation, N and P fertilizer
Results: LER

No fertilizer added:
MC Sorghum 1557kg/ha; uninoculated Chickpea 854kg/ha;
inoculated Chickpea 848kg/ha
Sorghum (1.11) and uninoculated Chickpea (0.14)   1.25
Sorghum (1.24) and inoculated Chickpea (0.14)   1.38

20kg/ha N added:
MC Sorghum 1919kg/ha; uninoculated Chickpea 924kg/ha;
inoculated Chickpea 788kg/ha
Sorghum (0.96) and uninoculated Chickpea (0.15)   1.11
Sorghum (1.17) and inoculated Chickpea (0.14)   1.31

20kg/ha N and 100kg/ha P added:
MC Sorghum 3430kg/ha; uninoculated Chickpea 814kg/ha;
inoculated Chickpea 1046kg/ha
Sorghum (1.00) and uninoculated Chickpea (0.11)   1.11
Sorghum (0.89) and inoculated Chickpea (0.07)   0.96

12. Crookston, R.K. (1976), Nigeria; USA
Intercrops used: Cotton, Cowpea
Results: LER
USA: Sorghum and Cotton   1.50
Nigeria: Sorghum and Cowpea   1.50

13. Gebrekidan, B. (1977), Ethiopia
Intercrops used: Bean
Results: LER
MC Sorghum 50q/ha; MC Bean 21q/ha
Sorghum (0.76) and Bean (0.95)   1.71

14. Mohta, N.K. and R. De (1980), New Delhi, India
Intercrops used: Soybean
Variables tested: density, planting pattern
Results: LER
MC Sorghum (rows 45cm apart) 2.00tons/ha
Alternate rows of Sorghum and Soybean with Soybean rows
    30cm apart   1.20
Alternate pairs of rows of Sorghum and Soybean with Soybean
    rows 30cm apart   1.28
Alternate rows of Sorghum and Soybean with Soybean rows
    22.5cm apart   1.12
Alternate rows of Sorghum and Soybean with Soybean rows
    45cm apart   1.31

15. Natarajan, M. and R.W. Willey (1980), Hyderabad, India
Intercrops used: Pigeon pea
Variables tested: density
Experimental parameters: different planting densities were used; fLER values were as high as 1.45

Results: LER
MC Sorghum 4467kg/ha; MC Pigeon pea 1 017kg/ha
Sorghum (0.93) and Pigeon pea (0.74) 1.67
Sorghum (0.97) and Pigeon pea (0.66) 1.63
Sorghum (0.92) and Pigeon pea (0.64) 1.56
Sorghum (0.97) and Pigeon pea (0.62) 1.59
Sorghum (0.90) and Pigeon pea (0.70) 1.60
Sorghum (0.95) and Pigeon pea (0.72) 1.67

16. Norman, D.W. (1974), Nigeria
Intercrops used: Cowpea, Millet, Peanut
Variables tested: four-crop combination
Results: LER
MC Sorghum 792kg/ha
* Sorghum and Cowpea and Millet and Peanut 1.66

17. Rao, M.R. and R.W. Willey (1980a), Hyderabad, India
Intercrops used: Castor bean, Cowpea, Foxtail millet, Maize, Peanut, Pearl millet, Pigeon pea, Soybean
Variables tested: soil type
Experimental parameters: the soils used were Alfisols, with a low water-holding capacity, and Vertisols, with a higher water-holding capacity
Results: LER

1975

Alfisols
Sorghum (0.80) and Cowpea (0.40) 1.20
Sorghum (0.75) and Foxtail millet (0.40) 1.15
Sorghum (0.50) and Pearl millet (0.65) 1.15
Sorghum (0.75) and Pigeon pea (0.45) 1.20
Sorghum (0.85) and Soybean (0.40) 1.25

Vertisols
Sorghum (0.85) and Cowpea (0.15) 1.00
Sorghum (0.75) and Foxtail millet (0.45) 1.20
Sorghum (0.55) and Pearl millet (0.70) 1.25
Sorghum (0.80) and Pigeon pea (0.75) 1.55
Sorghum (0.85) and Soybean (0.15) 1.00

1976

Alfisols
Sorghum (1.00) and Castor bean (0.30) 1.30
Sorghum (0.85) and Cowpea (0.50) 1.35
Sorghum (0.95) and Foxtail millet (0.35) 1.30
Sorghum (0.40) and Maize (0.90) 1.30
Sorghum (1.10) and Peanut (0.30) 1.40
Sorghum (0.80) and Pearl millet (0.65) 1.45

Vertisols
Sorghum (1.00) and Cowpea (0.35) 1.35
Sorghum (0.90) and Foxtail millet (0.65) 1.55
Sorghum (0.90) and Maize (0.80) 1.70

Sorghum (1.00) and Peanut (0.20)	1.20
Sorghum (0.95) and Pearl millet (0.90)	1.85

Because 1976 was a drier year than 1975, the water-holding capacity of Vertisols had a greater effect on crop yields

18. Rao, M.R. and Willey, R.W. (1980b)
Intercrops used: Pigeon pea
Experimental parameters: results from many experiments are summarized

Results:	LER
65 experiments: Sorghum and Pigeon pea	1.42
64 cases: 2 rows Sorghum and 1 row Pigeon pea	1.43
40 cases: 1 row Sorghum and 1 row Pigeon pea	1.40

19. Reddi, K.C.S. *et al.* (1980), Hyderabad, India
Intercrops used: Greengram, Pigeon pea
Variables tested: N fertilizer
Experimental parameters: experiments were done at 2 farms

Results:	LER	fLER
At Apau farm:		
0kg/ha N added: MC Sorghum 1820kg/ha		
Sorghum (0.89) and Greengram (0.23)	1.12	1.26
Sorghum (0.86) and Pigeon pea (0.62)	1.48	2.07
40kg/ha N added: MC Sorghum 2710kg/ha		
Sorghum (0.92) and Greengram (0.15)	1.07	1.15
Sorghum (0.90) and Pigeon pea (0.43)	1.33	1.75
80kg/ha N added: MC Sorghum 2953kg/ha		
Sorghum (0.94) and Greengram (0.13)	1.07	1.14
Sorghum (0.90) and Pigeon pea (0.38)	1.28	1.64
At ICRISAT farm:		
0kg/ha N added: MC Sorghum 1423kg/ha		
Sorghum (0.91) and Greengram (0.29)	1.20	1.35
Sorghum (0.90) and Pigeon pea (0.36)	1.26	1.61
40kg/ha N added: MC Sorghum 2437kg/ha		
Sorghum (0.99) and Greengram (0.17)	1.16	1.25
Sorghum (0.95) and Pigeon pea (0.20)	1.15	1.36
80kg/na N added: MC Sorghum 2747kg/ha		
Sorghum (1.00) and Greengram (0.12)	1.12	1.17
Sorghum (0.94) and Pigeon pea (0.17)	1.11	1.28

20. Reddy, K.A. *et al.* (1980), Hyderabad, India
Intercrops used: Blackgram, Cowpea, Finger millet, Foxtail millet, Greengram, Peanut, Soybean, Sunflower
Variables tested:
Experimental parameters: paired rows of Sorghum alternating with two rows of the intercrop were compared to normal planting of Sorghum

Results:	LER	fLER
Sorghum and Blackgram	1.50	1.54
Sorghum and Cowpea	1.51	1.52
Sorghum and Finger millet	1.26	1.24
Sorghum and Foxtail millet	1.28	1.25

Sorghum and Greengram	1.49	1.50
Sorghum and Peanut	1.60	1.70
Sorghum and Soybean	1.30	1.32
Sorghum and Sunflower	1.01	1.15

21. Rego, T.J. (1979), Hyderabad, India
Intercrops used: Pigeon pea
Variables tested: N fertilizer
Experimental parameters: Sorghum and Pigeon pea were intercropped in alternate rows with 45cm between the rows; N fertilizer was added to Sorghum
Results: LER
MC Pigeon pea 1390kg/ha

0kg/ha N added: MC Sorghum 930kg/ha
Sorghum (1.10) and Pigeon pea (0.58)   1.68

60kg/ha N added: MC Sorghum 2890kg/ha
Sorghum (0.80) and Pigeon pea (0.59)   1.39

120kg/ha N added: MC Sorghum 4590kg/ha
Sorghum (0.84) and Pigeon pea (0.63)   1.47

22. Satyanarayana, D.V. *et al.* (1979), Hyderabad, India
Intercrops used: Cowpea, Peanut, Safflower, Soybean, Sunflower, Wheat
Experimental parameters: Sorghum rows were 60cm apart
Results:

MC Sorghum grain 45.6q/ha; MC Sorghum Rs4228/ha

Intercrop	Sorghum grain yield (q/ha)	IC Yield (q/ha)	LER	fLER
Cowpea	39.8	5.6	1.00	1.07
Peanut	44.5	4.6	1.08	1.25
Safflower	44.3	3.0	1.04	1.09
Soybean	51.0	6.2	1.25	1.50
Sunflower	24.1	18.1	0.93	1.23
Wheat	46.1	5.1	1.12	1.17

23. Shetty, S.V.R. and A.N. Rao (1979), Hyderabad, India
Intercrops used: Pigeon pea
Variables tested: density, proportion of intercrops
Results: LER
MC Sorghum 4043kg/ha – medium density, 180 000 pl/ha
MC Pigeon pea 1704kg/ha – medium density, 40 000 pl/ha
Sorghum 90 000pl/ha (0.52) and Pigeon pea 20 000pl/ha (0.47) 0.99
Sorghum 90 000pl/ha (0.60) and Pigeon pea 40 000pl/ha (0.57) 1.17
Sorghum 90 000pl/ha (0.63) and Pigeon pea 80 000pl/ha (0.59) 1.22
Sorghum 180 000pl/ha (0.72) and Pigeon pea 20 000pl/ha
  (0.47)   1.19
Sorghum 180 000pl/ha (0.65) and Pigeon pea 40 000pl/ha
  (0.62)   1.27
Sorghum 180 000pl/ha (0.72) and Pigeon pea 80 000pl/ha
  (0.81)   1.53

Sorghum 360 000pl/ha (0.66) and Pigeon pea 20 000pl/ha
(0.39) 1.05
Sorghum 360 000pl/ha (0.78) and Pigeon pea 40 000pl/ha
(0.76) 1.54
Sorghum 360 000pl/ha (0.77) and Pigeon pea 80 000pl/ha
(0.63) 1.40

24. Singh, S.P. (1981), New Delhi, India
Intercrops used: Blackgram, Cowpea (fodder), Cowpea (grain), Greengram, Peanut
Variables tested: density, planting pattern
Experimental parameters: paired rows of Sorghum were alternated with 1 or 2 rows of the intercrop; results are an average of 2 years: 1975 and 1976.

Results:

	A	B	C	D	
Distance between Sorghum rows (cm):	30	30	30	60	
Space left for intercrop (cm):	60	60	90	60	
# rows intercrop planted in space:	1	2	2	1	
					Ave. LER
Sorghum and Blackgram LERs:	1.68	1.86	1.93	1.91	1.85
Sorghum and Cowpea (fodder) LERs:	1.69	1.80	1.89	1.84	1.80
Sorghum and Cowpea (grain) LERs:	1.67	1.87	2.19	1.94	1.92
Sorghum and Greengram LERs:	1.61	1.78	1.98	1.91	1.82
Sorghum and Peanut LERs:	1.37	1.47	1.52	1.47	1.46
Average LERs:	1.51	1.63	1.75	1.68	

25. Tarhalkar, P.P. and N.G.P. Rao (1979), Hyderabad, India
Intercrops used: Castor bean, Greengram, Peanut, Pigeon pea, Sunflower, Soybean
Variables tested: density, Pigeon pea varieties, planting pattern

Results:

	LER	fLER
1971 Kharif		
MC Sorghum 4125kg/ha, value Rs3217/ha		
Sorghum and Castor bean	1.08	1.10
Sorghum and Greengram	1.06	0.96
Sorghum and Peanut	1.57	1.00
Sorghum and Pigeon pea	1.04	0.81
Sorghum and Soybean	1.61	1.29
Sorghum and Sunflower	0.88	0.95
1973 Kharif:		
MC Sorghum 6770kg/ha, value Rs6772/ha		
MC Sorghum paired rows 7571kg/ha, value Rs7571/ha		
Sorghum and Castor bean	1.29	1.12
Sorghum and Peanut	1.26	1.11
Sorghum and Pigeon pea	1.54	1.24
Sorghum and Soybean	1.43	1.15

1974 Kharif:

	LERs	
	Paired rows 60–30cm	Wide rows 90cm
Sorghum and Castor bean	1.11	1.19
Sorghum and Peanut	1.28	1.35
Sorghum and Pigeon pea	1.43	1.45
Sorghum and Soybean	1.42	1.58

1975 Kharif:
Sorghum and bushy 130-day Pigeon pea (HY-2)

	LERs	
	Pigeon pea	
Sorghum planting pattern	27 000pl/ha	55 000pl/ha
Paired rows (60-30cm)	1.71	1.29
Paired rows (90-30cm)	1.51	1.35
Wide rows (60cm)	1.38	1.32
Wide rows (90cm)	1.51	1.59

1976 Kharif:
Sorghum and Pigeon pea

	LERs				
Sorghum planting pattern	HY-3A Pigeon pea (bushy) Density		HY-2 Pigeon pea (erect) Density		
	Low	High	Low	High	Average
Paired rows (60–30cm)	1.46	1.35	1.33	1.23	1.34
Paired rows (90–30cm)	1.48	1.40	1.34	1.37	1.40
Wide rows (60cm)	1.46	1.44	1.35	1.38	1.41
Wide rows (90cm)	1.40	1.47	1.44	1.48	1.45
Average	1.45	1.42	1.37	1.37	

26. Willey, R.W. (1979b), various locations
Intercrops used: Pigeon pea
Variables tested: N fertilizer

Results:	LER	fLER
0kg/ha N added: Sorghum and Pigeon pea	1.46	
40kg/ha N added: Sorghum and Pigeon pea	1.52	
80kg/ha N added: Sorghum and Pigeon pea	1.38	
120kg/ha N added: Sorghum and Pigeon pea	1.46	
120kg/ha N added compared to no N added: Sorghum and Pigeon pea		1.79

Soybean

1. Brown, C.M. and D.W. Graffis (1976), Illinois, USA
Intercrops used: Oat (grain), Oat (silage)

Results:	LER
MC Soybean 136bu/ha; MC Oat (silage) 4056kg/ha	
Soybean (0.72) and Oat (silage) (1.05)	1.77
Soybean 69bu/ha (0.51) and Oat (grain) 234bu/ha (0.87)	1.38

2. Chatterjee, B.N. and M.A. Roquib (1975), West Bengal, India
Intercrops used: Maize, Pigeon pea, Rice
Experimental parameters: results are an average of 2 years; cLERs were calculated in terms of calories of yield

Results:	LER	fLER	cLER
MC Soybean 2876kg/ha, Rs4301/ha, 124 kilocalories/ha			
Soybean and Maize	1.58	1.41	1.32
Soybean and Pigeon pea	1.46	1.07	1.28
Soybean and Rice	1.51	1.29	1.37

3. Chatterjee, B.N. and M.A. Roquib (1975), West Bengal, India
Intercrops used: Maize
Variables tested: N fertilizer
Experimental parameters: results are an average of 2 years

Results:	LER
20kg/ha N added:	
Soybean 535kg/ha (0.35) and Maize 1755kg/ha (1.59)	1.94
100kg/ha N added:	
Soybean 470kg/ha (0.31) and Maize 2670kg/ha (0.90)	1.21

4. Finlay, R.C. (1974), Tanzania
Intercrops used: Cowpea, maize, Sorghum
Variables tested: planting pattern

Results:	LER
MC Soybean 17.7kg/ha; MC Cowpea 10.0kg/ha; MC Maize 5660kg/ha; MC Sorghum 3880kg/ha	
Same row: Soybean and Maize	1.31
Alternate rows: Soybean and Maize	1.30
Alternate 2 rows: Soybean and Maize	1.05
Same row: Soybean and Sorghum	1.15
Alternate rows: Soybean and Sorghum	1.08
Alternate 2 rows: Soybean and Sorghum	1.08
Same row: Cowpea and Maize	1.00
Same row: Cowpea and Sorghum	0.89
Same row: Maize and Sorghum	1.05
Alternate rows: Maize and Sorghum	0.92

Sugar

1. Bhadauria, V.A. and B.K. Mathur (1973), Uttar Pradesh, India
Intercrops used: Sannhemp
Experimental parameters: Sannhemp was used as green manure

Results:	LER	fLER
MC Sugar 46 500kg/ha		
Sugar and Sannhemmp	1.05	1.04

Appendix I: Land equivalent ratios

2. Bose, P.K. and H. Ashraf (1972), Bihar, India
Intercrops used: Maize, Mustard, Pea, Potato, Wheat
Variables tested: proportion of intercrops
Experimental parameters: 90cm rows were used
Results:

Proportion	Combination	LER	fLER
1:1	Sugar and Maize	0.74	1.03
1:3	Sugar and Mustard	0.69	0.77
1:3	Sugar and Pea	0.76	0.76
1:1	Sugar and Potato	1.18	1.55
1:3	Sugar and Wheat	0.67	1.20

3. Chatterjee, B.N. *et al.* (1978), West Bengal, India
Intercrops used: Potato
Results: LER   fLER
MC Sugar cane (average) 1446q/ha, Rs16 778/ha net
MC Potato (average) 267q/ha
Sugar (0.91) and Potato (0.81)   1.72   1.20
Sugar and Potato compared to MC Potato   3.65

4. Deol, D.S. (1978), Punjab, India
Intercrops used: Sugar beet
Experimental parameters: results are an average of 6 experiments
Results: LER   fLER
MC Sugar 103.5tons/ha; MC Sugar beet 59tons/ha
Sugar and Sugar beet   1.67   1.79

5. Kar, K.R. and J.S.S. Dixit (1975), Uttar Pradesh, India
Intercrops used: Lahi, Potato, Wheat
Results: fLER
MC Autumn sugar 119tons/ha
Sugar and Lahi   1.06
Sugar and Potato   1.37
Sugar and Wheat   0.84

6. Krutman, S. (1968), Brazil
Intercrops used: Cowpea, Kidney bean
Variables tested: planting pattern
Results: LER
MC Sugar 141tons/ha; MC Cowpea 1350kg/ha
Sugar (1.00) and Cowpea (1.00)   2.00
MC Sugar 140tons/ha; MC Kidney bean 1190kg/ha
Mixed within the row: Sugar and Kidney bean   1.32
Bean row 20cm from Sugar row: Sugar and Kidney bean   1.47
Bean row halfway between Sugar rows: Sugar and Kidney bean   1.65

7. Mathur, B.K. (1976), Uttar Pradesh, India
Intercrops used: Potato, Wheat
Results: fLER
MC Sugar Rs4976/ha
Sugar and Potato   1.34
Sugar and Wheat   1.12

8. Ramdhawa, K.S. (1975), Haryana, India
Intercrops used: Capsicum, Chilli, Okra, Onion, Tomato
Results:

	fLER
MC Sugar net income Rs3215/ha	
Sugar and Capsicum	2.57
Sugar and Chilli	2.58
Sugar and Okra	2.58
Sugar and Onion	3.95
Sugar and Tomato	2.95

9. Rathi, K.S. and R.A. Singh (1979), various places, India
Intercrops used: Mustard, Potato
Results:

MC Sugar (average profit) Rs5771/ha
Sugar and Mustard (average profit) Rs8145/ha
Sugar and Mustard fLERs: 0.96, 0.95, 1.02, 1.07, 1.09, 1.12, 1.34, 1.61, 1.66, 1.37, 1.90, 1.97, 1.75, 1.81, 1.18, 1.06, 0.95, 0.92, 1.44

MC Sugar (average profit) Rs3949/ha
Sugar and Mustard (average profit) Rs4710/ha
Sugar and Mustard fLERs: 1.17, 1.18, 1.20, 1.20, 1.22, 1.20

MC Sugar (average profit) Rs6680/ha
Sugar and Potato (average profit) Rs9445/ha
Sugar and Potato fLERs: 1.35, 1.35, 1.36, 1.68, 1.58, 1.37, 1.28, 1.34, 1.20, 1.49, 0.54

MC Sugar (average profit) Rs4845/ha
Sugar and Potato (average profit) Rs6457/ha
Sugar and Potato fLERs: 1.20, 1.26, 1.26, 1.84, 1.24

10. Sharma, R.A. (1979), Madhya Pradesh, India
Intercrops used: Okra, Onion, Potato, Sugar beet, Wheat
Variables tested: density, two- and three-crop combinations
Results:

	LER	fLER
Sugar rows 90cm apart:		
MC Sugar 787q/ha, Rs4591/ha (net return)		
* Sugar and Okra and Potato	1.19	1.27
Sugar and Onion	1.29	1.89
* Sugar and Okra and Wheat	0.63	0.39
Sugar and Sugar beet	1.14	0.53
Sugar rows 60cm apart:		
MC Sugar 655.5q/ha, Rs2621/ha (net return)		
* Sugar and Okra and Potato	1.03	0.77
Sugar and Onion	1.27	2.29
* Sugar and Okra and Wheat	0.76	0.71
Sugar and Sugar beet	1.19	0.29

11. Singh, P.P. and A. Singh (1974), Uttar Pradesh, India
Intercrops used: Wheat
Variables tested: spacing, Wheat varieties
Results: fLER
MC Sugar Rs6857/ha
Sugar and Wheat 1.14
Sugar and Wheat 1.20
Sugar and Wheat 1.38

Sunflower

1. Ahmed, S. and H.P.M. Gunasena (1979), Philippines
Intercrops used: Mung bean
Variables tested: N fertilizer
Results: LER fLER

% recommended dose of N fertilizer added	Combination		
100	Sunflower and Mung bean	1.95	2.08
50	Sunflower and Mung bean	1.53	1.54
0	Sunflower and Mung bean	1.20	1.44

2. Ahmed, S. *et al.* (1976), Philippines
Intercrops used: Mung bean
Variables tested: N fertilizer
Results: LER fLER

Sunflower and Mung bean:

N added	Total yield	Total value		
120kg/ha	2301kg/ha	Rs727.28/ha	1.96	4.41
60kg/ha	1613kg/ha	Rs451.43/ha	1.49	2.41
0kg/ha	1374kg/ha	Rs507.76/ha	1.31	1.99

3. Shanthamalliah, N.R. *et al.* (1978), Mysore, India
Intercrops used: Bean, Cowpea, Field bean, Finger millet, Peanut, Soybean
Results: LER
MC Sunflower (average of 2 years) 7.64q/ha
Sunflower and Bean 1.14
Sunflower (0.55) and Cowpea 3.54q/ha (0.69) 1.24
Sunflower (0.62) and Field bean 3.07q/ha (0.56) 1.18
Sunflower (0.43) and Finger millet 7.94q/ha (0.48) 0.91
Sunflower (0.76) and Peanut 3.64q/ha (0.46) 1.22
Sunflower (0.71) and Soybean 4.29q/ha (0.53) 1.24

4. Singh, K.C. and R.D. Singh (1977), Rajasthan, India
Intercrops used: Cluster bean, Cowpea, Greengram, Mothbean, Peanut
Variables tested: N fertilizer, proportion of intercrops
Results: LER fLER
1 row Sunflower and 1 row Cluster bean 1.66
1 row Sunflower and 1 row Cowpea 1.80
1 row Sunflower and 1 row Greengram 1.64
1 row Sunflower and 1 row Mothbean 1.43
1 row Sunflower and 1 row Peanut 1.47

2 rows Sunflower and 1 row Cluster bean	1.98	1.03
2 rows Sunflower and 1 row Cowpea	2.16	1.63
2 rows Sunflower and 1 row Greengram	1.93	1.44
2 rows Sunflower and 1 row Mothbean	1.47	1.06
2 rows Sunflower and 1 row Peanut	1.71	1.43

Effect of N addition on LERs and fLERs:
1975 and 1976 results are combined:

No N added:
Sunflower and Cowpea	3.40	2.47
Sunflower and Greengram	3.78	2.53

30kg/ha N added:
Sunflower and Cowpea	2.98	2.34
Sunflower and Greengram	3.08	2.25

60kg/ha N added:
Sunflower and Cowpea	2.26	1.80
Sunflower and Greengram	3.25	1.71

Wheat

1. Bains, S.S. (1968), India
Intercrops used: Chickpea, Lentil, Pea
Variables tested: irrigation
Results: LER

Non-irrigated:
MC Wheat 1410kg/ha; MC Chickpea 1190kg/ha
Wheat and Chickpea   1.26
MC Wheat 9700kg/ha; MC Lentil 700kg/ha
Wheat and Lentil   2.35
MC Wheat 9600kg/ha; MC Pea 1550kg/ha
Wheat and Pea   1.38

Irrigated:
MC Wheat 2610kg/ha; MC Chickpea 2580kg/ha
Wheat and Chickpea   1.15

2. Chan, L.M. and C.M. Brown (1980), Illinois, USA
Intercrops used: Soybean
Variables tested: density
Results: LER

21cm between rows:
MC Wheat 3812kg/ha; MC Soybean 2559kg/ha
Wheat (1.05) and Soybean (0.11)   1.16

41cm between rows:
MC Wheat 2472kg/ha; MC Soybean 2785kg/ha
Wheat (1.12) and Soybean (0.35)   1.47

Appendix I: Land equivalent ratios

61cm between rows:
MC Wheat 1724kg/ha; MC Soybean 2644kg/ha
Wheat (1.07) and Soybean (0.93) 2.00

81cm between rows:
MC Wheat 1305kg/ha; MC Soybean 2888kg/ha
Wheat (1.08) and Soybean (0.77) 1.85

3. Chowdhury, S.L. (1979), Ranchi, India
Intercrops used: Linseed
Results: LER   fLER

MC Wheat 640kg/ha; MC Linseed 670kg/ha
Wheat (0.31) and Linseed (0.81)   1.16   1.40

MC Wheat 610kg/ha; MC Linseed 840kg/ha
Wheat (0.46) and Linseed (0.57)   1.03   1.11

4. Chowdhury, S.L. (1979), Jhansi, India
Intercrops used: Chickpea
Results: LER   fLER
MC Wheat 690kg/ha; MC Chickpea 470kg/ha
Wheat (1.08) and Chickpea (0.38)   1.46   1.30

5. Chowdhury, S.L. (1979), Samba, India
Intercrops used: Chickpea
Results: LER   fLER

MC Wheat 4990kg/ha; MC Chickpea 1240kg/ha
Wheat (0.78) and Chickpea (0.13)   0.91   0.92

MC Wheat 1920kg/ha; MC Chickpea 620kg/ha
Wheat (0.75) and Chickpea (0.39)   1.14   1.21

6. Jeffers, D.L. (1979), Ohio, USA
Intercrops used: Soybean
Variables tested: planting pattern, timing
Results: LER
MC Wheat (7in rows) 64bu/acre; MC Soybean (21in rows) 44bu/acre
Wheat 7in rows (0.84) and Soybean 21in rows (0.66)   1.50
Wheat 3 rows, skip 1 (0.73) and Soybean 28in rows (0.82) 1.55
Wheat 2 rows, skip 1 (0.75) and Soybean 21in rows (0.89) 1.64

Planted *Planting pattern*
April 29   Wheat 7in rows (0.77) and Soybean 21in rows (0.93)   1.70
May 9   Wheat 7in rows (0.70) and Soybean 21in rows (0.80)   1.50
May 16   Wheat 7in rows (0.69) and Soybean 21in rows (0.75)   1.44

Yam

1. Hegde, D.M. (1981), Karnataka, India
Intercrops used: Cluster bean, Cowpea, Kidney bean

Results:	LER	fLER
MC Yam 13 985kg/ha		
Medicinal Yam (0.95) and Cluster bean	1.29	1.30
Medicinal Yam (0.96) and Cowpea	1.33	1.41
Medicinal Yam (0.95) and Kidney bean	1.39	1.49

Appendix 2: Plant names

Amaranth	*Amaranthus*, L.
Anise	*Pimpinella Anisum*, L.
Asparagus	*Asparagus* spp.
Avocado	*Persea americana*, Mill.
Bajri (pearl millet)	*Pennisetum typhoideum*, Rich.
Balsam pear	*Momordica charantia*
Banana	*Musa* spp.
Barti (millet)	*Echinochloa Stagnina*, Beauv.
Beet	*Beta vulgaris*, L.
Blackgram	*Phaseolus Mungo*, L.
Black mustard	*Brassica nigra*, (L) Koch.
Breadfruit	*Artocarpus altilis (Park) Fosberg.*
Broadbean	*Phaseolus lunatus*, L.
Cabbage	*Brassica oleracea*, L.
Cacao	*Theobroma cacao*, L.
Calalu	*Xanthosoma hastifolium*
Carrot	*Daucus Carota*, L.
Cassava	*Manihot esculenta*, Crantz.
Castor-bean	*Ricinus communis*, L.
Cauliflower	*Brassica oleracea*
Cedar	*Cedrela odorata*, L.
Chayote	*Sechium edule*, Swartz.
Chickpea	*Cicer arietinum*, L.
Chilli	*Capsicum frutescens*, L.
Chocho	*Sechium edule*, Swartz
Citrus	*Citrus*, spp.
Cluster bean	*Cyamopsis tetragonaloba*, (L.) Taub.
Coconut	*Cocos nucifera*, L.
Coffee	*Coffee arabica*
Collard	*var. Acephala*, DC.
Coriander	*Coriandrum sativum*, L.
Corn	*Zea mays*, L.
Cowpea	*Vigna sinensis*, L.
Cucumber	*Cucumis sativus*
Deccan hemp	*Hibiscus cannabinus*, L.
Dill	*Anethum graveolens*, L.
Eggplant	*Solanum melongena*
Fennel	*Foeniculum vulgare*, Mill.
Fenugreek	*Trigonella foenum-graecum*
Finger millet	*Eleusine Coracana*, L.
Foxtail millet	*Setaria italica*
Garlic	*Allium sativum*
Ginger	*Zingiber officinale*
Gourd	*Lagenaria* spp.
Greengram	*Phaseolus aureus*

Horsegram	*Dolichos uniflorus*
Horseradish	*Armoracia lapathifolia*
Hyacinth bean	*Lablab niger*
Irish potato	*Solanum tuberosum*
Jowar	*Sorghum* spp.
Kidney bean	*Phaseolus vulgaris*, L.
Kohlrabi	*Brassica oleracea*
Lettuce	*Lactuca sativa*
Lima bean	*Phaseolus lunatus*
Linseed	*Linum usitatissimum*
Maize	*Zea mays*, L.
Mango	*Mangifera indica*
Millet	*Panicum millaceum*, L.
Moth bean	*Phaseolus aconitifolius*
Mustard	*Brassica juncea*
Niger seed	*Guizotia abyssinica*
Okra	*Hibiscus sabdariffa*
Onion	*Allium Cepa var. cepa*
Papaya	*Carica Papaya*
Pea	*Pisum sativum*
Peanut	*Arachis hypogaea*, L.
Pigeon-pea	*Cajanus Cajan*
Pimento	*Pimenta dioica*
Poppy	*Papaver* spp.
Proso millet	*Panicum miliaceum*, L.
Pumpkin	*Cucurbita* spp.
Radish	*Raphanus sativus*, L.
Rayo (broad-leafed mustard)	*Brassica juncea*
Rice	*Oryza sativa*
Safflower	*Carthamus tinctoria*
Sannhemp	*Crotolaria juncea*
Sesame	*Sesamum indicum*
Scallion	*Allium ascalonicum*
Sisal	*Agave sisalana*
Sorghum	*Sorghum bicolor*
Soybean	*Glycine Max* (L.) Merr.
Spinach	*Spinacia oleracea*, L.
Spurry	*Spergula arvensis*
Sugar (sugar cane)	*Saccharum cvs.*
Sweet potato	*Ipomoea batatas*
Tara	*Caesalpinia spinosa* (Mol) Ktze.
Taro	*Colocasia esculenta* (L.) Schott.
Tomato	*Lycopersicon esculentum*
Trumpet tree	*Cecropia peltata*
Turmeric	*Curcuma longa*
Turnip	*Brassica rapa*, L.
Wheat	*Triticum aestivum*
Yam	*Dioscorea* spp.

Bibliography

Abraham, K.J. (1974) 'Intercropping in Arecanut Helps to Build Up Farmer's Economy', *Arecanut and Spices*, Vol. 5 (3), March: 73–75.

Abraham, P.P. (1973) 'Smallholders Project Research: Intercropping', *Rubber Research Institute of Malaysia Annual Report*: 172–85.

Adesiyun, A.A. (1979) 'Effects of Intercrop on Fruit Fly, Oscinella Frit, Oviposition and Larval Survival on Oats', *Entomologica Experimentalis et Applicata*, Vol. 26: 219–22.

Agboola, A.A. and A.A. Fayemi (1971) 'Effect of Interplanted Legumes and Fertilizer Treatment on the Major Soil Nutrients', *International Symposium on Soil Fertility Evaluation Proceedings*, Vol. 1: 529–40.

Ahmed, S. and H.P.M. Gunasena (1979) 'N Utilization and Economics of Some Intercropped Systems in Tropical Countries', *Tropical Agriculture*, Vol. 56 (2), April: 115–23.

Ahmed, S., M. Sadiq and L. Kobayashi (1976) 'Studies on Mixed Intercropping Sunflower with Mung. Proceedings First Review Meeting INPUTS Project', Central Luzon State *University Scientific Journal*: 259–68, Nuevo Ecija, Philippines.

Ahn, P.M. (1993) *Tropical Soils and Fertiliser Use*, London: Longman.

Akhanda, A.M., Jose T. Mauco, V.E. Green and G.M. Prine (1978) 'Relay Intercropping Peanut, Soybean, Sweetpotato and Pigeonpea in Corn', *Soil and Crop Science Society of Florida*, Proceedings, Vol. 37: 95–101.

Alexander, M.W. and C.F. Gentner (1962) 'Production of Corn and Soybeans in Alternate Pairs of Rows', *Agronomy Journal*, Vol. 54 (3): 233-4.

Anderson, E. and L.O. Williams (1954) 'Maize and Sorghum as a Mixed Crop in Honduras', *St Louis, Missouri Botanical Garden Annals*, Vol. 41 (2): 213–15.

Anderson, T. (1976) 'Subsystems of *Conuco* Agriculture in the Basin of Lake Valencia, Venezuela: A Classification and Description', *Ohio Geographers: Recent Research Themes*, Vol. 4: 24–34.

Andrews, D.J. (1972) 'Intercropping with Sorghum', *International Symposium on Sorghum in the 1970's*, New Delhi: Oxford & IBH Publishing Co., Vol. 6: 546–55.

Anthony, K.R.M. and S.G. Willimott (1957) 'Cotton Interplanting Experiments in the south-west Sudan', *Empire Journal of Experimental Agriculture*, Vol. 25: 29–36.

Appadurai, R. and R.K.V. Selva (1974) 'A Note on Groundnut–Redgram Mixture in Lower Bhavani Project Area', *Madras Agricultural Journal*, Vol. 61 (9 pt. II): 803–4.

Arangzeb, S.N.M. (1966) 'Intercropping of Comilla Cotton', *Pakistan Cottons*, Vol. 10 (4): 172–6.

Attems, M. (1968) 'Permanent Cropping in the Usambara Mountains: The Relevance of the Minimum Benefit Thesis', in H. Ruthenberg (ed.) *Smallholder Farming and Smallholder Development in Tanzania*, Afrika Studien No. 24, Münich: Weltforum Verlag, IFO: 137–74.

Ayyengar, G.N.R. and M.A. Sankara Ayyar (1941) 'Rotation and Mixed Crops with Sorghum', *Madras Agricultural Journal*, Vol. 29: 57–63.

Baeumer, K. and C.T. DeWit (1968) 'Competitive Interference of Plant Species in Monoculture and Mixed Stands', *Netherlands Journal of Agricultural Science*, Vol. 16 (2): 103–22.

Bains, S.S. (1968) 'Pulses are Popular for Mixed Cropping', *Indian Farming*, Vol. 17 (11): 21–22.

Baker, E.F.I. (1974) 'Research on Mixed Cropping with Cereals in Nitrogen Farming Systems – A System for Improvement', mimeograph, International Workshop on Farming Systems, ICRISAT, International Crops Research Institute for the Semi-arid Tropics, 1–11–256 Begumpet, Hyderabad, 500016 (AD), India.

Baker, E.F.I. (1978) 'Mixed Cropping in Northern Nigeria. I. Cereals and Groundnuts', *Expl. Agric.*, Vol. 14: 293–8.
Baker, E.F.I. (1979a) 'Mixed Cropping in Northern Nigeria. III. Mixtures of Cereals', *Expl. Agric.*, Vol. 15: 41–8.
Baker, E.F.I. (1979b) 'Population, Time and Crop Mixtures', prepared for International Intercropping Workshop, ICRISAT, Hyderabad, India.
Baker, E.F.I. and the Institute for Agricultural Research (IAR), Ahmadu Bello University, Nigeria (1979) 'Population, Time and Crop Mixtures', mimeograph, prepared for the International Intercropping Workshop, 10–13 January ICRISAT, Hyderabad, India.
Baker, E.F.I. and Y. Yusuf (n.d.) 'Research with Mixed Crops at the Institute for Agricultural Research, Samaru, Nigeria'.
Bangham, W.N. (1946) 'Una Excelente Combinacion Para los Tropicos Americanos', Caracas: Editorial Elite/Publication of Comite Ejecutiva (93rd) Tercera Conferencia Interamericana de Agricultura: 5–21.
Barney, G. (1980) *The Global 2000 Report to the President of the US Entering the 21st Century*, New York: Pergamon Press.
Bartlett, C.D.S. and G.I. Mlay (1976) *Assessment of Innovations in Inter-cropping Systems*, Symposium on Inter-cropping in Semi-arid Areas, Morogoro, University of Dar-es-Salaam, Faculty of Agriculture.
Beckerman, S. (1983b) 'Does the Swidden Ape the Jungle?' *Human Ecology* 11 (1): 1–12.
Beets, W.C. (1975) 'Multiple-cropping Practices in Asia and the Far East', *Agriculture and Environment*, Vol. 2 (3), October: 219–28.
Beets, W.C. (1982) *Multiple Cropping and Tropical Farming Systems*, Aldershot: Gower, 156 pp.
Belshaw, D. (1979) 'Taking Indigenous Technology Seriously: The Case of Intercropping Techniques in East Africa', *IDS Bulletin* 10 (2): 24–7.
Belshaw, D. (1980) 'Taking Indigenous Knowledge Seriously: The Case of Intercropping Techniques in East-Africa', in D.W. Brokensha, D.M. Warren and O. Werner (eds) *Indigenous Knowledge Systems and Development*, Lanham: University Press of America, Inc.: 195–203.
Bengtsson, A. (1973) 'Forsok med Samodling av Arter och Havre/Experiments with Joint Cultivation of Peas and Oats (Intercropping)', *Uppsala. Lantbrukshoegskolan och Statens Lantbruksfoersoek. MeddeLanden. Serie A*, (English Title: *Reports*), Vol. 200, summer, 24 pp.
Benjamini, L. (1980) 'Bait Crops and Mesurol Sprays to Reduce Bird Damage to Sprouting Sugar Beets', *Phytoparasitica*, Vol. 8 (3): 151–61.
Beste, C.E. (1976) 'Co-cropping Sweet Corn and Soybeans', *Hortscience*, Vol. 11 (3): June 236–8.
Bhadauria, V.A. and B.K. Mathur (1973)'Problem of Green Manuring Sugarcane – Intercropping as a Solution', *Indian Sugar*, Vol. 23 (4), July: 351–8.
Bhalerao, S.S. (1976) 'Intercropping Studies in Sorghum', *Sorghum Newsletter*, Vol. 19: 63–4.
Bhatawadekar, P.U., S.S. Chiney and K.M. Deshmukh (1966) 'Response of Bajri-Tur Mixture Crop to Nitrogen and Phosphate Under Dry Farming Conditions of Sholapur', *Indian Journal of Agronomy*, Vol. 11 (3): 243–6.
Bhatnagar, V.S. and J.C. Davies (1979) 'Pest Management in Intercrop Subsistence Farming, *Cropping Entomologist & Cereal Entomologist*, January 1979, prepared for International Intercropping Workshop, ICRISAT, Hyderabad, India.
Borse, R.H., U.B. Mahajan and S.H. Shinde (1980) 'Studies on Companion Cropping with Sorghum (Sorghum Bicolor Moench)', *Geobios*, Vol. 7: 36–8.
Bose, P.K. and H. Ashraf (1972) 'Intercropping in Autumn Planted Sugarcane in North Bihar and the Economics Involved', *Indian Sugar (Calcultta)*, Vol. 22 (6): 507–13.
Braud, M. and F. Richez (1964) 'Sur la Possibilite d'un Premier Cycle Vivriere avant une

Production Cotonniere en Second Cycle dans la Region de Bambari (Centrafrique)', *Cotons et Fibres Tropicales*, Vol. 18 (3), (French edition).
Bray, F., (1994) 'Agriculture for Developing Nations', *Scientific American*, July: 18–25.
Brokensha, D.W. (1990) 'Local Management Systems and Sustainability', in C.H. Gladwin and K. Truman (eds) *Food and Farm: Current Debates and Policies*, Monographs in Economic Anthropology No. 7, Lanham: University Press of America.
Brougham, R.W. (1958) 'Interception of Light by the Foliage of Pure and Mixed Stands of Pasture Plants', *Australian Journal of Agricultural Research*, Vol. 9 (1): 39–52.
Brown, C.M. and D.W. Graffis (1976) 'Intercropping Soybeans and Sorghums in Oats', *Illinois Research*, Vol. 18 (2), Spring: 3–4.
Buranday, R.P. et al. (1975) 'Effects of Cabbage–Tomato Intercropping on the Incidence and . . . ', *Philippine Entomologist*, July: 369–74.
Carvalho, Mario de (1969) 'O Algodao Como Cultura Intercalar do Sisal', *Revista Agricola* (Associacao do Fomento Agricola e Industrial de Mocambique), Vol. 11 (108): 7–8.
Chambers, R., A. Pacey and L.A. Thrupp (eds) (1989) *Farmer First: Farmer Innovation and Agricultural Research*, London: Intermediate Technology Publications, 240 pp.
Chambers, R. and B.P. Ghildyal (1985) 'Agricultural Research for Resource-poor Farmers: The Farmer-First-and-Last Model', Institute of Development Studies No. DP203, 29 pp. Brighton, University of Sussex.
Chan, L.M. and C.M. Brown (1980) 'Relay Intercropping Soybeans into Winter Wheat and Spring Oats', *Agronomy Journal*, Vol. 72 (1), Jan/Feb, Madison, Wisconsin.
Chanda, R. (1988) 'The Human and Environmental Factors in Traditional Crop Complex Development in Zambia', *Singapore Journal of Tropical Geography* 9 (1): 33–44.
Chandler, P. (1994) 'Shamu Jianzhong: A Traditionally Derived Understanding of Agroforestry Sustainability in China', *Journal of Sustainable Forestry* 1 (4): 1–24.
Chandrasekharan, N.R. et al., (1974) 'Mixed Cropping with Sesamum', *Madras Agricultural Journal*, Vol. 61 (8): 510–15.
Chandravanshi, B.R. (1975) 'Study on Intercropping in Sorghum (Sorghum bicolor (L.) Moench) under Uniform and Paired Row Planting Systems', Jawaharlal Nehru Krishi Vishwa Vidyalaya, *JNKVV Research Journal*, Vol. 9 (1/2), Jan/Apr.
Chatterjee, B.N., N.C. Banerjee and D.C. Ghosh (1978) 'Fertilizer Application in Intercropping of Sugar Cane and Wheat', *Fertilizer News*, New Delhi, Vol. 23 (10), October.
Chatterjee, B.N. and M.A. Roquib (1975) 'Soybean in Multiple and Intercropping', *Indian Journal of Genetics and Plant Breeding*, Vol. 35, July: 264–8.
Cheng, T.S. (1979) 'A New Approach to Interplanting Maize with Sugarcane in Taiwan', *Taiwan Sugar*, Vol. 26 (3), May/June: 95–7.
Chengappa, P.G. and N.S.P. Rebello (1977) 'An Economic Analysis of Intercropping in Coffee Estates of . . . ', *South Indian Horticulture*, Vol. 25 (4), October: 154–7.
Chew, W.Y. (1979) *Intercropping with Cassava*, IDRC, Ottawa, Canada, 'Cassava Intercropping Patterns and Practices in Malaysia', Malaysian Agriculture Research and Development Institute, Jalan Kebun, Kelong, Selangor, Malaysia, in Weber and Campbell (eds) (1979).
Chowdhury, M.S. and R.N. Misangu (1979) Faculty of Agriculture, Forestry and Veterinary Science, University of Dar es Salaam, PO Box 643, Morogoro, Tanzania, 'Sorghum-Chickpea Intercropping Trial at Morogoro, Tanzania', mimeograph, prepared for the International Intercropping Workshop, 10–13 January, ICRISAT, Hyderabad, India.
Chowdhury, S.L. (1979) 'Recent Studies in Intercropping Systems on the Drylands of India – Some Thoughts, Some Results', mimeograph, prepared for the International Intercropping Workshop, 10–13 January, ICRISAT, Hyderabad, India.
Christianty, L., O.S. Abdoellah, G.G. Marten et al. (1986) 'Traditional Agroforestry in West Java: The Pekarangan (Homegarden) and Kebun-Talun (Annual–Perennial Rotation) Cropping Systems', in G.G. Marten (ed.) *Traditional Agriculture in South-East Asia: A Human Ecology Perspective*, London/Boulder CO: Westview Press: 132–58.

Cordero, A. and R.E. McCollum (1979) 'Yield Potential of Interplanted Annual Food Crops in Southeastern US', *Agronomy Journal*, Vol. 71, Sept./Oct: 834–42.
Cox, Jeff (1979) 'Companion Planting for Harmony and Production', *Organic Gardening*, February: 56–64.
Crookston, R.K. (1976) 'Intercropping: A New Version of an Old Idea', *Crops and Soils*, Vol. 28 (9), Aug/Sept: 77-9.
Cummins, D.G. (1973) 'Interplanting of Corn, Sorghum and Soybeans for Silage', *Georgia, Agricultural Experiment Stations, Research Bulletin*, Vol. 150, December, 15 pp.
Dalal, R.C. (1974) 'Effects of Intercropping Maize with Pigeon Peas on Grain Yield and Nutrient Uptake', *Experimental Agriculture*, Vol. 10: 219–24.
Dalal, R.C. (1977) 'Effect of Intercropping of Maize with Soya Bean on Grain Yield', *Tropical Agriculture*, Vol. 54 (2), April: 189–91.
Daulay, H.S. (1978) 'Intercropping System for Drylands', in *Arid Zone Research and Development* (ed.) H.S. Mann: 229–37.
Daulay, H.S. *et al.* (1970) 'Intercropping of Grasses and Legumes', *Indian Farming*, Vol. 19 (10), January: 12–14, 44.
De, Rajat and S.P. Singh (1979) 'Management Practices for Intercropping Systems', prepared for International Intercropping Workshop, ICRISAT, Hyderabad, India, January.
De, Rajat and R.B.L. Bhardwaj (1974) 'Systems of Multiple Cropping to Maximize Economic Returns and Employment', Rome: Food and Agriculture Organization of the United Nations/First FAO/SIDA Seminar on Improvement and Production of Field Food Crops for Plant Scientists from Africa and the Near East.
Dempster, J.P. and T.H. Coaker (1972) 'Diversification of Crop Ecosystems as a Means of Controlling Pests', *Biology in Pest and Disease Control* (eds) J.D. Price and M.E. Solomon, New York: John Wiley & Sons, 13th Symposium of the British Ecological Society, 4–7 January.
Deol, D.S. (1978) 'Effects of Plant Population on the Yield of Sugar Beet . . . , *International Sugar Journal*, July: 355–7.
Desai, N.D. and S.N. Goyal (1980) 'Intercropping of Sesame with other Oilseed Crops', *Indian Journal of Agricultural Science*, Vol. 50: 603–5.
Devos, P. and G.F. Wilson (1979) 'Intercropping of Plantains with Food Crops: Maize, Cassava and Cocoyams', *Fruits*, Vol. 34 (3): 169–73.
Devotta, A.D. and S.R. Chowdappan (1975) 'A Note on Mixed Cropping Dryland Cotton (Bavathi)', *Madras Agricultural Journal*, Vol. 62 (4): 234–6.
Divekar, C.B. and F.B. Kurtakoti (1961) 'Studies Relating to the Intercropping of Groundnut in Cotton', *Indian Cotton Growing Review*, Vol. 15 (4): 233ff.
Dogra, B. (1983) 'Traditional Agriculture in India: High Yields and No Waste', *Ecologist* 13 (2–3): 84–7.
Eagleton, G.E., A.A. Mohamed, A.A. Odowa *et al.* (1991) 'A Comparison of Moisture-conserving Practices for the Traditional Sorghum-based Cropping Systems of the Bay Region in Somalia', *Agriculture, Eco-Systems and Environment* 36 (1–2): 87–99.
Edwards, R. (1987a) 'Mapping and Informal Experimentation by Farmers: Agronomic Monitoring of Farmers' Cropping Systems', IDS Paper, Workshop on Farmers and Agricultural Research: Complementary Methods, 12 pp.
Effendi, S. (n.d.) 'Cassava Intercropping Patterns and Management Practices in Indonesia', Bogor, Indonesia: Central Research Institute for Agriculture.
Ehrenfeld, D. (1978) *The Arrogance of Humanism*, NY: Oxford Press.
Enyi, B.A.C. (1973) 'Effects of Intercropping Maize or Sorghum with Cowpeas, Pigeon Peas or Beans', *Experimental Agriculture*, Vol. 9 (1), January: 83–90.
Evans, A.C. (1960) 'Studies of Intercropping I', *East African Agricultural and Forestry Journal* 26: 1–10.
FAO Report (1956) 'Agriculture Nomade en Côte d'Ivoire; the Dranouas Tribe in the Interior', Rome.
Fernandes, E.C.M., A. O'kting'ati and J. Maghembe (1984) 'The Chagga Homegardens:

A Multistoried Agroforestry Cropping System on Mount Kilimanjaro, Northern Tanzania', *Agroforestry Systems* 2 (2): 73–86.
Finlay, R.C. (1974) 'Intercropping Soybeans with Cereals', *Soybean Production, Protection, and Utilization, Proceedings, Addis Ababa*, Urbana-Champaign: College of Agriculture, University of Illinois, INTSOY Series Number 6.
Fisher, N.M. (1976) 'Experiments with Maize–Bean and Maize–Potato Mixed Crops in an Area with Two Short Rainy Seasons in the Highlands of Kenya', Nairobi, Kenya: Department of Crop Science, University of Nairobi.
Fisher, N.M. (1977) 'Studies in Mixed Cropping. I. Seasonal Differences in Relative Productivity of Crop Mixtures and Pure Stands in the Kenya Highlands', *Expl. Agric.*, Vol. 13: 177–84.
Fisher, N.M. (1979a) 'Polyculture Cropping Systems of Africa and their Implications for Appropriate Tillage Technologies', Zaria, Nigeria: Agronomy Section, Institute for Agricultural Research, Ahmadu Bello University, January.
Fisher, N.M. (1979b) 'The Improvement of Grain Legumes for Mixed Cropping Systems', Zaria, Nigeria: Institute for Agricultural Research, Ahmadu Bello University, August.
Fisher, N.M. (1979c) 'Studies in Mixed Cropping. III. Further Results with Maize–Bean Mixtures', *Expl. Agric.*, Vol. 15: 49–58.
Francis, C.A. (ed) (1986) *Multiple Cropping Systems*, New York: Macmillan, 383 pp.
Francis, C.A. (1992) 'Sustainability Issues with Intercrops' in W. Hiemstra *et al.* (ed.) *Let Farmers Judge: Experiences in Assessing the Sustainability of Agriculture*, London: Intermediate Technology Publications.
Francis, C.A., M. Prager and G. Tejada (1982a) 'Effects of Relative Planting Dates in Bean (Phaseolus Vulgaris L.) and Maize (Zea Mays L.) Intercropping Patterns', *Field Crops Research*, Amsterdam, Vol. 5 (1), March: 45–54.
Francis, C.A., M. Prager and G. Tejada (1982b) 'Density Interactions in Tropical Intercropping II. Maize (Zea Mays L.) and Bush Beans (Phaseolus Vulgaris L.)', *Field Crops Research*, Amsterdam, Vol. 5 (3), August: 153–264.
Gahlot, K.N.S. *et al.* (1978) 'Urd "T.9" as Intercrop with Ahar "T.21"', *Indian Farming*, Vol. 27: 7–8.
Gangwar, Babooji and G.S. Kalkar (1982) 'Intercropping Rainfed Maize with Different Legumes', *Indian Journal of Agricultural Science*, New Delhi, Vol. 52 (2), February: 113–16.
Gebrekidan, Brhane (1977) 'Sorghum–Legume Intercropping in the Chercher Highlands of Ethiopia', *AAASA Journal*, Vol. 4 (2), July.
Genest, J. and H. Steppler (1973) 'Effects of Companion Crops and their Management on the Undersown Forage Seedling Environment', *Canadian Journal of Plant Sciences*, Vol. 53: 285–90.
Godoy, R.A. and P.A. Bennett (1991) 'The Economics of Monocropping and Intercropping by Smallholders: The Case of Coconut in Indonesia', *Human Ecology* 19 (1): 83–98.
Gold, C.S. (1993) 'Effects of Cassava Intercropping and Varietal Mixtures on Herbivore Load, Plant Growth, and Yields: Applications for Small Farmers in Latin America', in M. Altieri (ed.) *Crop-protection Strategies for Subsistence Farmers*, London: Intermediate Technology Publications.
Goonatilake, Soosantha (1984) *Aborted Discovery; Science and Creativity in the Third World*, Zed Books, London.
Graves, C.R. (1978) 'Double Cropping System Using Soybeans or Grain Sorghum and Wheat', *Tennessee Farm and Home Science Progress Report*, Vol. 106: 4–6.
Greenland, D.J. (1975) 'Bringing the Green Revolution to the Shifting Cultivator', *Science*, Vol. 190 (4217), November: 841–4.
Grimes, R.C. (1963) 'Intercropping and Alternative Cropping of Maize and Cotton', *East African Agriculture and Forestry Journal* 28: 161–3.
Gunasena, H.P.M. (1980) 'Performance of Maize–Legume Intercrop Systems in Sri Lanka', Tanzania: 2nd SISA Conference.

Gupta, G.P. and B.P. Mathur (1964) 'Mixed Cropping Studies in Oilseeds', *Indian Oilseeds Journal*, Vol. 8: 206–13.
Gurung, B.D. (1993) 'The Use of Fallow Land: Prospects for Sustainable Agricultural Production', *Farm Management Notes for Asia and the Far East* 16: 17–27.
Haizel, K.A. (1972) 'The Productivity of Mixtures of Two and Three Species', *Journal of Applied Ecology*, Vol. 9 (2): 601–8.
Hardjono, J. (ed.) (1991) *Indonesia: Resources, Ecology, and Environment*, Oxford: Oxford University Press, 280 pp.
Hardjono, J. (1993) *Land, Labour and Livelihood in A West Java Village*, Yokyakarta: Gadjah Mada University Press, 323 pp.
Hasselbach, O.E. and A.M.M. Ndegwa (1980) 'Modifying the Competitive Relationship in Maize–Bean Mixture in Kenya', Tanzania: 2nd SISA Conference, August.
Hegde, D.M. (1981) 'Intercropping in Medicinal Yam with Short-duration Cowpea, Clusterbean and Kidney-bean', *Indian Journal of Agricultural Science*, Vol. 51 (4): 262–5, Indian Institute of Horticultural Research, Bangalore, Karnataka 560 080.
Hegde, D.M. and C.S. Saraf (1979) 'Economics of Phosphorus Fertilization of Intercropping Systems in Pigeon Pea under Dryland Conditions', *Fertilizer News*, New Delhi, Vol. 24 (3), March: 28–30, 40.
Herrara, W.A.T., B.T. Samson and R.R. Hardwood (1979) 'The Effect of Row Arrangement and Plant Density on Productivity of Corn–Rice and Corn–Peanut Intercrops', *Philippine Journal of Crop Science*, Vol. 1 (3), NST: 125–8.
Hunter, J.R. and E. Camacho (1961) 'Some Observations on Permanent Mixed Cropping in the Humid Tropics', *Turrialba*, Vol. 11 (1): 26–33.
Institut de Recherches Agronomiques Tropicales (IRAT) (1969) 'Cameroun-Addendum du Rapport Annuel 1968: Compte-Rendu de la Fertilisation des Cultures Associees (Dschang, Foumbot)', *IRAT, Rapport Analytique*, Vol. I: 131–57.
IRAT (1965) Institute de Recherches Agronomiques Tropicales, Haute-Volta, #358, 'Essai de Culture en Melange Niebe-Cotonnier', in *IRAT-Haute-Volta, Rapport Annuel*: 62–3.
IRAT (1975) Institute de Recherches Agronomiques Tropicales, Mission IRAT au Togo, #359, 'Cultures Associees', in *Experimentation Agronomique dans les Regions Maritimes et des Savanes*: 31–2.
IRAT (1978) *Les Cultures Associees Bibliographie 1977*, Paris: Institut de Recherches Agronomique Tropicales.
Jackson, Wes (1980) *New Roots for Agriculture*, Friends of the Earth, San Francisco.
Jain, T.C. and G.N. Rao (1980) 'Note on a New Approach to Analysis of Yield Data in Intercropping System', *Indian J. Agric. Sci.*, Vol. 50 (12), December: 570–2.
Jeffers, D.L. (1979) 'Management Needed for Relay Intercropping Soybeans and Wheat', *Ohio Report on Research and Development; Agriculture, Home Economics and Natural Resources*, Vol. 64 (5), Sept/Oct: 67–70.
Jodha, N.S. (1977) 'Resource Base as a Determinant of Cropping Patterns', *ICRISAT Occasional Paper 14*, Hyderabad, India: International Crops Research Institute for the Semi-Arid Tropics, April.
Jodha, N.S. (1979a) 'Some Dimensions of Traditional Farming in Semi-arid Tropical India', International Crops Research Institute for the Semi-arid Tropics, Progress Report No. 4.
Jodha, N.S. (1979b) *Intercropping in Traditional Farming Systems: Andrah Pradesh India*, Andhra Pradesh: International Crops Research Institute for the Semi Arid Tropics (ICRISAT), 25 pp.
Jodha, N.S. (1980) 'Some Dimensions of Traditional Farming Systems in Semi-arid Tropical India', in *Socio-Economic Constraints to Development of Semi-arid Tropical Agriculture*, Proceedings International Workshop, Hyderabad, India, 19–23 February 1979, Patancheru, Andhra Pradesh: International Crops Research Institute for the Semi-arid Tropics: 11–24.
Johnson, R. and D.J. Allen (1975) 'Induced Resistance to Rust Diseases and its Possible Role in the Resistance of Multiline Varieties', *Ann. Appl. Biol.*, Vol. 80: 359–63.

Joshi, S.N. and H.U. Joshi (1965) 'Mixed Cropping of Groundnut–Cotton under Irrigated Condition in Saurashtra', *Indian Oilseed Journal*, Vol. 9 (4): 244–8.
Juo, A.S.R. and R. Lal (1977) 'The Effect of Fallow and Continuous Cultivation on the Chemical and Physical Properties of an Alfisol in Western Nigeria', *Plant and Soil*, Vol. 47 (3), August: 567–84.
Kairon, M.S. and D.S. Nandal (1971) 'Economics of Intercropping of Mung and Cowpeas in Cotton', *Allahabad Farmer*, Vol. 45 (2), March: 239–41.
Kanogo, T. (1992) 'Women and Environment in History' in S.A. Khasiani (ed.) *Groundwork: African Women as Environmental Managers*, Nairobi: ACTS Press: 7–19.
Kar, K.R.S. and J.S.S. Dixit (1975) 'Intercropping with Autumn-planted Sugarcane in the Tarai Tract of Uttar Pradesh', *Indian Sugar*, Vol. 25 (1), April: 27–30.
Kaul, J.N. and H.S. Sikhon (1974) 'Intercropping Studies with Arhar (Cajanus Cajan) (Pigeonpeas)', *Journal of Research*, Punjab Agricultural University, Ludhiana, India, Vol. 11 (2), June: 158–63.
Kaushik, R.D. (1951) 'Mixed Cropping in the Delhi State', *Allahabad Farmer*, Vol. 25 (4): 142–9.
Kayumbo, H.Y. and S.J. Asman (1980) 'Energy Studies for Rural Tanzania: The Case for Intercropping in Dodoma', Morogoro, Tanzania: 2nd SISA Conference, August.
Keswani, C.L. and R.A.D. Mreta (1980) 'Effect of Intercropping on the Severity of Powdery Mildew on Greengram', Morogoro, Tanzania: 2nd SISA Conference, August.
Khader, K.B.A. and K.J. Anthony (1968) 'Intercropping: A Paying Proposition for Area Growers – What Crops to Grow', *Indian Farming*, Vol. 18 (4): 14–15.
Khehra, A.S., H.S. Brar and R.K. Sharma (1979) 'Studies on Inter-cropping of Maize (Zea Mays L.) with Black Gram (Phaseolus Mungo Roxb.)', *Indian J. Agric. Res.*, Vol. 13 (1): 23–6.
King, C. et al. (1978) 'Interplanting Corn and Soybeans Yields', *Highlights of Agricultural Research*, Vol. 25 (1), Spring: 6ff.
King, K.F.s. and M.T. Chandler (1978) 'The Wasted Lands', *The Programme of Work of the International Council for Research in Agroforestry*, Nairobi, Kenya, September: 1–35.
King, F.H. (1911) *Farmers of Forty Centuries*, Rodale Press, Emmaus, PA.
Koregrave, B.A. (1964) 'Effect of Mixed Cropping on the Growth and Yield of Suran (Elephant Yam, *Amorphophallus campanulatus*)', *Indian Journal of Agronomy*, Vol. 9 (4): 255–60.
Kotschi, J., A. Waters Bayer et al. (1989) *Ecofarming in Agricultural Development*, No. 2, Weikersheim, Margraf: Tropical Agroecology, GTZ, 143 pp.
Krantz, B.A. (1979) Agronomist and Program Leader, Farming Systems Research Program, ICRISAT, 'Intercropping on an Operational Scale in an Improved Farming System', mimeograph, prepared for the International Intercropping Workshop, 10–13 January, ICRISAT, Hyderabad, India.
Krings, T. (1991) 'Indigenous Agricultural Development and Strategies for Coping with Famine the Case of Senoufo (Pomporo) in Southern Mali (West Africa)', *Bayreuther Geowissenschaftliche Arbeiten Naturwissenschaftliche Gesellschaft Bayreuth*, 15: 69–81.
Krutman, S. (1968) 'Cultura Consorciada Cana X Feijoeiro. Primeiros Resultados', *Pesquisa Agropecuaria Brasileira*, Vol. 3: 127–34.
Kubsad, S.C. and V.S. Dasaraddi (1974) 'Multiple Cropping with Hybrid Cotton in Sindhanur Taluk', *Farm Front*, Vol. 8 (8), August: 6–9.
Kurtz, T. (1952) 'The Importance of Nitrogen and Water in Reducing Competition Between Intercrops and Corn', *Agronomy Journal*, Vol. 44: 13–17.
Kwitny, Jonathan (1984) *Endless Enemies. The Making of an Unfriendly World*, Congdon and Weed, New York.
Lacsina, R. and S.K. De Datta (1975) 'Integrated Weed Management Practices for Controlling a Difficult Weed in Lowland Rice', *Philippine Weed Science Bulletin*, Vol. 2 (1/2): 1–5.

Lado, C. (1988) 'A Note on "Traditional" Agricultural Land Use: A Case Study from Maridi District, Southern Sudan', *Malaysian Journal of Tropical Geography* 18: 25–34.

Lado, C. (1989) 'Traditional Agricultural Land Use and Some Changing Trends in Maridi District, Southern Sudan', *Land Use Policy* 6 (4): 324–40.

Lal, R. (1975) 'Soil Erosion Problems on an Alfisol in Western Nigeria and Their Control', Ibadan, Nigeria: International Institute for Tropical Agriculture (IITA).

Lappé, Frances Moore and Joseph Collins (1978) *Food First*, Ballantine, New York.

Laycock, D.H. and R.A. Wood (1963) 'Some Observations on Soil Moisture Use Under Tea in Myasaland. III. The Effect of Cover . . . ', *Tropical Agriculture (Trinidad)*, Vol. 40 (2): 121–8.

Lee, S.A. (1972) 'Agro-Economic Studies on Intercropping in Pineapple', *Malaysian Pineapple*, Vol. 2: 23–32.

Liboon, S.P., R.P. Harwood and H.G. Zandstra (1976) 'The Effect of Crop Damage in Corn–Peanut Intercropping', Davao City, Paper presented at the 7th Annual Scientific Meeting of the Crop Science Society of the Philippines, 10–12 May.

Lipman, J.G. (1912) 'The Associative Growth of Legumes and Nonlegumes', *New Jersey Agricultural Experiment Station, New Brunswick Bulletin No. 253*: 3–48.

Malithano, A.D. and J. Van Leeuwen (n.d.) 'Groundnut–Maize Interplanting in Southern Mozambique', Maputo, Mozambique: Faculty of Agriculture and Forestry, University Eduardo Mondlane.

Martin, M.P.L.D. and R.W. Snaydon (1982) 'Intercropping Barley and Beans: I. Effects of Planting Pattern', *Experimental Agriculture*, Cambridge, Vol. 18 (2), April: 139–40.

Mathews, R.B., S.T. Holden, J. Volk *et al.* (1992) 'The Potential of Alley Cropping in Improvement of Cultivation Systems in the High Rainfall Areas of Zambia, I: Chitimene and Fundukila', *Agroforestry Systems* 17 (3): 219–40.

Mathur, B.K. (1976) 'Increasing Agricultural Production through Intercropping in Autumn Planted Sugarcane', *Cane Grower's Bulletin*, July: 5–10.

Matthews, D. (1981) 'The Effectiveness of Selected Herbs and Flowers in Repelling Garden Insect Pests', Rodale Press Inc.: Organic Gardening and Farming Research Center, 21 pp.

May, K.W. and R. Misangu (1980) 'Some Observations of the Effects of Plant Arrangements for Intercropping', in *Proceedings of the Second Symposium of Intercropping for Semi-Arid Areas* held at the Faculty of Agriculture, Forestry and Veterinary Science, University of Dar es Salaam, Morogoro, Tanzania, 4–7 August 1980, IDRC, Ottawa, Canada.

Mead, R. and R.W. Willey (1980) 'The Concept of a "Land Equivalent Ratio" and Advantages in Yields from Intercropping', *Experimental Agriculture*, Vol. 16 (3), July: 217–28.

Mohamed, S.A. (1989) *Problems of Agricultural Production Sustainability and Fuel Wood Supply in Rural Tanzania: Can Agroforestry Help?* Research Paper No. 23, Dar es Salaam: Institute of Resource Assessment, University of Dar es Salaam, 33 pp.

Mohta, N.K. and R. De (1980) 'Intercropping Maize and Sorghum . . . ', *Journal of Agricultural Science*, Vol. 95 (11), August: 117–22.

Mollison, Bill (1979) *Permaculture One and Two*, Tagari, Australia.

Moody, K. and S.V.R. Shetty (1979) 'Weed Management in Intercropping Systems', prepared for the International Intercropping Workshop, ICRISAT, Hyderabad, India, January.

Moreno, R.A. (n.d.) 'Crop Protection Implications of Cassava Intercropping', Turrialba, Costa Rica: Centro Agronomico Tropical de Investigacion y Ensenanza.

Moreno, Raul, A. and R.D. Hart (1979) 'Intercropping with Cassava in Central America', in Weber and Campbell (eds) (1979).

Mugabe, N.R., M.E. Singe and K.P. Sibuga (1980) 'A Study of Crop/Weed Competition in Intercropping', Morogoro, Tanzania: 2nd SISA Conference, August.

Mukiibi, J.K. (n.d.) 'Effect of Intercropping on Some Diseases of Beans and Groundnuts', Kampala, Uganda: Department of Crop Science, Makerere University.

Murray, G.F. (1986) 'Seeing the Forest while Planting the Trees: An Anthropological Approach to Agroforestry in Rural Haiti', in D.W. Brinkerhoff and J.C. García-Zamar (eds) *Politics, Projects and People: Institutional Development in Haiti*, New York: Praeger: 193–226.

Nadar, H.M. and G.E. Rodewald (1979) 'Relay Cropping and Intercropping: An Approach to Maximize Maize Yield in the Marginal Rainfall Areas of Kenya', mimeograph, prepared for the International Intercropping Workshop, 10–13 January, ICRISAT, Hyderabad, India.

Nair, P.K.R. (1977) 'Pattern of Light Interception by Canopies in a Coconut Cacao Crop', *Indian Journal of Agricultural Sciences*, July: 453–62.

Natarajan, M. and R.W. Willey (1980) 'Sorghum–Pigeonpea Intercropping and the Effects of Plant Population Density', *Journal of Agricultural Science*, Vol. 95 (1), August: 59–65.

Netting, R., M.P. Stone and G.D. Stone (1989) 'Kofyar Cash Cropping: Choice and Change in Indigenous Agricultural Development', *Human Ecology* 17 (3): 299–319.

Nichol, H. (1935) 'Mixed Cropping in Primitive Agriculture', *Empire Journal of Experimental Agriculture*, 3: 189–95.

Norman, D.W. (1974a) 'Crop Mixtures under Indigenous Conditions in the Northern Part of Nigeria', *Samaru Research Bulletin*, Vol. 205, reprinted from *Factors of Agricultural Growth in West Africa*, Institute of Social and Economic Research, Ghana, 1973: 130–44.

Norman, D.W. (1974b) 'An Economic Survey of Three Villages in Zaria Province', *Samaru Miscellaneous Paper 37*, Zaria, Nigeria: Institute for Agricultural Research, Ahmadu Bello University.

Norman, D.W. (1974c) 'Rationalizing Mixed Cropping under Indigenous Conditions: The Example of Northern Nigeria', *Journal of Development Studies*, Vol. 11, October: 3–21.

Norman, D.W. (1977a) 'Economic Rationality of Traditional Hausa Dryland Farmers in North of Nigeria', in R. Stevens (ed.) *Traditions and Dynamic in Small Farm Agriculture: Economic Studies in Asia, Africa and Latin America*, Ames: Iowa State University Press: 63–91.

Norman, D.W. (1977b) 'The Rationalisation of Intercropping', *African Environment*, July: 97–109.

Nuru, S. (1984) 'The Adjustment and Integration of Livestock and Cropping Systems', in D.L. Hawksworth (ed.) *Advancing Agricultural Production in Africa*, Proceedings of CAB's 1st. Scientific Conference, Arusha, Tanzania: Slough, CAB: 257–62.

Nweke, F. *et al.* (1980) 'Bases for Farm Resources Allocation in the Smallholders Cropping Systems of Nigeria: A Case Study of Awka and Abakaliki Villages', Ibadan, IITA, Discussion Paper, 480 pp.

Odhiambo, T.R. (1990) 'Assets of an IPM Specialist with Particular Reference to Chilo;, *Insect Science Application* 11 (4–5): 571–6.

Oelsligle, D.D., R. Menenses and R.E. McCollom (1976) 'Nitrogen Response by a Corn–Cassava Intercropped System in the Atlantic Coast of Costa Rica', *Agron. Econ. Res. Trop. Soils Ann. Rep.*, North Carolina, State University, Raleigh, Dept. of Soil Science, 1975: 197–200.

Ojomo, O.A. (1976) 'Development of Cowpea (*Vigna unculculata*) Ideotypes for Farming Systems in Western Nigeria', Paper presented at the Symposium on Mixed Cropping for Semi Arid Areas, University of Dar-es-Salaam, Tanzania, 10–11 May.

Okigbo, B.N. and D.J. Greenland (1976) 'Intercropping Systems in Tropical Africa', in R.J. Papendick, P.A. Sanchez and G.B. Triplett (eds) *Multiple Cropping*, Proceedings of a Symposium, Madison, ASA.

Okigbo, B.N. and R. Lal (1977) 'Role of Cover Crops in Soil and Water Conservation', *Food and Agriculture Organization of the United Nations, Soils Bulletin*, Vol. 33: 97–108.

Okorji, E.C. and C.O.B. Obiechina (1985) 'Bases for Farm Resource Allocation in the Traditional Farming System: A Comparative Study of Productivity of Farm Resources in Abakaliki Area of Anambra State, Nigeria', *Agricultural Systems* 17 (4): 197–210.

Olasantan, F.O. (1992) 'Vegetable Production in Traditional Farming Systems in Nigeria', *Outlook on Agriculture* 21 (2): 117–27.

Osiru, D.S.O. and G.R. Kibira (1979) 'Sorghum/Pigeonpea and Finger Millet/Groundnut Mixtures with Special Reference to Plant Population and Crop Arrangement', prepared for International Intercropping Workshop, ICRISAT, Hyderabad, India, January.

Patel, P.K. *et al.* (1979) 'Intercropping Groundnut and Soybean with Cotton', *Gujarat Agricultural University Research Journal*, Vol. 4 (2), January: 4–7.

Pauling, L. *et al.* (1983) 'Food/Forage Resource Base in a Traditional Rice Cropping System', 67th Annual Meeting, Federation of American Societies for Experimental Biology, Abstract No. 1455.

Pendleton, J.W., C.D. Bolen and R.D. Seif (1963) 'Alternating Strips of Corn and Soybeans versus Solid Plantings', *Agronomy Journal*, Vol. 55 (3): 293–5.

Pillai, M.R. *et al.* (1957) 'Mixed Cropping Trials with Ragi, Cotton and Groundnut', *Madras Agricultural Journal*, Vol. 44 (49): 131-9.

Pillai, P.N. (1974) 'Intercropping in Rubber', *Farming Facts (London)*, Vol. 8 (4), February: 29–31.

Pimentel, D. and M. (1977) 'Counting the Kilocalories: The Analysis of the Energetics of an Agricultural System Reveals the Immense Waste of Energy of an Economy Based on Profit', *Ceres*, 17: 20–21, Sept/Oct.

Polthanee, A. and G.G. Marten (1986) 'Rainfed Cropping Systems in Northeast Thailand', in G.G. Marten (ed.) *Traditional Agriculture in South East Asia: A Human Perspective*, Boulder CO/London: Westview Press: 103–31.

Porto, M.C.M., P.A. De Almeida, P.L.P. De Mattos and R.F. Souza (1979) 'Cassava Intercropping in Brazil', *Intercropping with Cassava*, Weber and Campbell (eds) (1979) Ottawa: IDRC.

Posner, J.L. (1982) 'Cropping Systems and Soil Conservation in the Hill Areas of Tropical America', *Turrialba* 32 (3): 287-99.

Puleston, D.E. (1978) 'Terracing, Raised Fields and Tree Cropping in the Maya Lowlands: A New Perspective in the Geography of Power', in P.D. Harrison and B.L. Turner (eds) *Pre-Hispanic Maya Agriculture*, Albuquerque: University of New Mexico Press: 225–45.

Pushparajah, E. and W.P. Weng (1970) 'Cultivation of Groundnuts and Maize as Intercrops in Rubber', *Conference on Crop Diversification in Malaysia – Proceedings*, E.K. and J.W. Blencoe (eds), Incorporated Society of Planters, PO Box 262, Kuala Lumpur, Malaysia.

Radke, J.K. (1968) 'Corn a Profitable Windbreak for Soybeans?', *Soybean Digest*, Vol. 28 (7): 17.

Raikhelkar, S.V. and U.C. Upadhyay (1980) 'Economics of Companion in Pearl Millet on Rainfed Lands', *Indian Journal of Agricultural Science*, Vol. 50 (9): 671–4.

Ramadasan, K., I. Abdullah, K.C. Tech and T. Vanialingam (1978) 'Intercropping of Coconuts with Cocoa in Malaysia', *Planter* (Kuala Lumpur), Vol. 54: 329–42.

Ramakrishnan, Nayar (1976) 'Intercropping in Young Robusta Coffee', *Indian Coffee*, Vol. 40 (2/3), Feb/Mar: 70–4.

Ramakrishnan, P.S. (1992) 'Ecology, Environment and Sustainable Rural Development in India', in C.R. Kartik *et al.* (eds) *Economic Development and Environment: A Case Study of India*, Delhi: Oxford University Press: 42–70.

Ramdhawa, K.S. (1975) 'Vegetables as Suitable Intercrops with Sugarcane', *Haryana J. Hort. Sci.*, Vol. 4 (3 and 4): 226–9.

Ranasinghe, D.M.S.H.K. and S.M. Newman (1993) 'Agroforestry Research and Practice in Sri Lanka', *Agroforestry Systems* 22 (2): 119–30.

Rao, M.R. and R.W. Willey (1980a) 'Preliminary Studies of Intercropping Combinations

Based on Pigeonpea or Sorghum', *Experimental Agriculture*, Vol. 16 (11), January: 29–39.
Rao, M.R. and R.W. Willey (1980b) 'Evaluation of Yield Stability in Intercropping Studies on Sorghum/Pigeonpea', *Experimental Agriculture*, Vol. 16 (2), April: 105–16.
Rathi, K.S. and R.A. Singh (1979) 'Companion Cropping with Autumn-planted Sugarcane – A Critical Review. II. Intercropping of Potato with Autumn-planted Sugarcane', *Indian Sugar Crops Journal*, Azad University of Agriculture and Technology, Kanpur, Sahibabad, India, Oct/Dec: 71 ff.
Rathi, K.S. and R.A. Singh (1979) 'Companion Cropping with Autumn-planted Sugarcane – A Critical Review. III. Intercropping of Mustard with Autumn-planted Sugarcane', *Indian Sugar Crops Journal*, C.S. Azad University of Agriculture and Technology, Kanpur, Sahibabad, India, Oct/Dec: 76–82.
Raynolds, M.K. and R.W. Elias (1980) 'The Agricultural Ecology of Intercropping', *Virginia Journal of Science*, Vol. 31 (4): 85.
Reddi, K.C.S., M.M. Hussain and B.A. Krantz (1980) 'Effect of Nitrogen Level and Spacing on Sorghum Intercropped with Pigeonpea and Greengram in Semi-arid Lands', *Indian Journal of Agricultural Science*, Vol. 50 (1): 17–22.
Reddy, G.J. and M.R. Reddi (1981) 'Studies on Intercropping in Maize Under Varied Row Spacings', *Indian Journal of Agronomy* (New Delhi), Vol. 26 (3): 360–2, September.
Reddy, K.A., K.R. Reddy and M.D. Reddy (1980) 'Effects of Intercropping on Yield and Returns in Corn and Sorghum', *Expl. Agric.*, Vol. 16: 179–84.
Reddy, M.H. and B.N. Chatterjee (1973) 'Intercropping of Soybean with Rice', *Indian Journal of Agronomy*, Vol. 18 (4), December: 464–72.
Reddy, M.S. and R.W. Willey (1979) 'Study of Pearl Millet/Groundnut Intercropping with Particular Emphasis on the Efficiencies of Leaf Canopy and Rooting Pattern', mimeograph, prepared for the International Intercropping Workshop, 10–13 January 1979, ICRISAT, Hyderabad, India.
Reddy, M.S. and R.W. Willey (1980) 'The Relative Importance of Above- and Belowground Resource Use in Determining Yield Advantages in Pearl Millet/Groundnut Intercropping', Tanzania: 2nd SISA Conference, August 1980.
Reddy, R.P. and P.P. Tarhalkar (1977) 'Improved Pigeonpea Varieties for Mono- and Intercropping', *Indian Farming*, Vol. 27 (9), December: 3–5, 35.
Rego, T.J. (1979) 'Nitrogen Response Studies of Intercropped Sorghum with Pigeonpea', mimeograph, prepared for the International Intercropping Workshop, 10–13 January 1979, ICRISAT, Hyderabad, India.
Reijntjes, C., B. Haverkort and A. Waters-Bayer (1992) *Farming for the Future: An Introduction to Low-External-Input and Sustainable Agriculture*, London: Macmillan Press Ltd., 250 pp.
Remison, S.U. (1980) 'Interaction between Maize and Cow Pea at Various Frequencies Intercropping', *Journal of Agricultural Science*, Cambridge Press, Vol. 94 (3), June: 617–21.
Richards, Paul (1985) *Indigenous Agricultural Revolution. Ecology and Food Procution in West Africa*, Hutchinson, Westview, London, Boulder, CO.
Richards, P., L.J. Slikkerveer and A.O. Phillips (1989) *Indigenous Knowledge Systems for Agriculture and Rural Development: The CIKARD Inaugural Lectures*, Studies in Technology and Social Change No. 13, Ames, Iowa State University, 40 pp.
Risch, S. (1980) 'The Population Dynamics of Several Herbivorous Beetles in a Tropical Agroecosystem: the Effect of Intercropping Corn, Beans and Squash in Costa Rica', *Journal of Applied Ecology*, Vol. 17 (3): 593–611, December.
Rodale, R. (1983) 'Pesticide Snap-Back', *Organic Gardening*, March: 30.
Rotenhan, D. von (1968) 'Cotton Farming in Sukumaland: Cash Cropping and its Implications', in H. Ruthenberg (ed.) *Smallholder Farming and Smallholder Development in Tanzania*, Afrika Studien No. 24, München: Weltforum Verlag, IFO: 51–86.

Roy, R.P., H.M. Sharma and H.C. Thakur (1981) 'Studies on Intercropping in Long Duration Pigeonpea on Sandy Loam Soil in North Bihar', *Indian Journal of Agronomy*, New Delhi, Vol. 26 (1), March: 77–82.

Sadanandan, N. and I.C. Mahapatra (1974) 'Influence of Multiple Cropping on the Water Stable Aggregates of Upland Rice Soils', *Agricultural Research Journal of Kerala*, Vol. 12 (1): 14–18, March.

Sampson, R.N. (1981) *Farmland or Wasteland: A Time to Choose*, Emmaus, Pa.: Rodale Press.

Saraf, C.S. (1975) 'Studies of Intercropping of Compatible Crops with Pigeon Pea', *Indian Journal of Agronomy*, Vol. 20 (2), June: 127–30.

Satyanarayana, D.V. et al. (1979) 'Studies on Intercropping in Grain Sorghum', *Indian Journal of Agronomy*, New Delhi, Vol. 24 (2), June: 223–4.

Sauer, C.O. (1952) 'Agricultural Origins and Dispersals', (Bowman Memorial Lectures, Series 2) New York: American Geographical Society.

Scherr, S.J. and P.A. Oduol (1989) *Alley Cropping and Tree Borders in Crop Fields in Siaya District, Kenya*, Nairobi, ICRAF and CARE International Collaborative Study, 24 pp.

Scherr, S.J., J.H. Roger and P.A. Oduol (1990) 'Surveying Farmers' Agroforestry Plots: Experiences in Evaluating Alley Cropping and Tree Border Technologies in Western Kenya', *Agroforestry Systems* 11 (2): 141–74.

Scoones, I. and J. Thompson (eds) (1994) *Beyond Farmer First: Rural People's Knowledge, Agricultural Research and Extension Practice*, London: Intermediate Technology Publications, 301 pp.

Searle, P.G., Y. Comudom, D. Sheddon and R.A. Nance (1981) 'Effect of Maize and Legume Intercropping Systems and Fertilizer Nitrogen on Crop Yields and Residual Nitrogen', *Field Crops Research*, Elsevier Scientific Pub. Co., Amsterdam, Vol. 4: 133–45.

Sebastiani, M. (1981) 'The Conucos Laguneros of the Lake Valencia Basin of Venezuela: an Appropriate Technology', Paper presented at I.G.U. Meetings, Fresno, California, April.

Sebasigari, K. (1985) 'Overview of Banana Cultivation and Constraints in the Economic Community of the Great Lakes States (CEPGL)', in R.A. Kirby and D. Ngendahayo (eds) *Banana Production and Research in Eastern and Central Africa*, Proceedings of a Regional Workshop in Bujumbura Burundi, Ottawa: Institut de Recherche Agronomique et Zootechnique: 9–22.

Shanthamallaiah, N.R., S. Purushotham and K.M. Krishnapp (1978) 'Studies of Intercropping with Sunflower', *Mysore Journal of Agricultural Sciences*, Vol. 12 (1): 41–4.

Sharma, K.N., D.S. Rana, S.R. Bishnoi and J.S. Sodhi (1979) 'Effect of Fertilizer Application in an Intercropping System', *Indian J. Agric. Res.*, Vol. 13 (1): 47–50.

Sharma, R.A. (1979) 'Intercropping with Sugarcane Pays More Profit', *Farmer and Parliament*, Vol. 14 (3), March: 11–12.

Shetty, S.V.R. and A.N. Rao (1979) 'Weed Management Studies in Sorghum/Pigeonpeas and Pearl Millet/Groundnut Intercrop Systems – Some Observations', prepared for International Intercropping Workshop, ICRISAT, Hyderabad, India, January.

Shiva, V., (1996) 'Towards A Biodiversity-Based Productivity Framework', *ILEIA Newsletter*, Vol. 12 (3), December: 22–23.

Simon, F. (1954) 'El Sembrio de Maiz Intercalado en los Algondonales: Experiencias en el Valle do Ate para el Control del *Heliothis virescens*', *Vida Agricola*, Vol. 31: 293–306.

Singh, C.M. and P. Chand (1980) 'Note on Economics of Grain Legume Intercropping and Nitrogen Fertilization in Maize', *Indian Journal of Agricultural Research*, Karnal, India, Vol. 14 (1): March: 62–4.

Singh, C.M. and W.S. Guleria (1979) 'Effect of Intercropping and Fertility Levels on Growth, Department and Yield of Maize', *Food Farming and Agriculture*, Vol. 10 (7), January: 242–4.

Singh, H.P. (1982) 'Intercropping with Arhar Pays Rich Dividends', *Indian Farming*, New Delhi, Vol. 32 (1), April: 19–20.
Singh, K.C. and R.D. Singh (1977) 'Intercropping of Annual Grain Legumes with Sunflower', *Indian Journal of Agricultural Science*, Vol. 47 (11): 563–7.
Singh, K.M. and R.N. Singh (1977) 'Succession of Insect Pests in Green Gram and Black Gram', *Indian Journal of Entomology*, Vol. 39 (4): 365–70, December.
Singh, Kalyan and M. Singh (1981) 'Intercropping of Urd with Arhar', *Indian Farming*, New Delhi, February: 11–12.
Singh, Kalyan, M. Singh, S.N. Shukal and B.N.R. Kushawaha (1978) 'Effect of Intercrops and Row Spacings on Growth and Yield of Pigeon Pea', *Indian Journal of Agricultural Research*, Vol. 12 (1): 27–31.
Singh, Mahatim, K. Singh, R.K. Singh and B.S. Saumitra (1979) 'Effect of Intercrops on the Growth and Yield of Pigeonpea', *Indian Journal of Agricultural Science*, Vol. 49 (2): 100–104.
Singh, P.P. and A. Singh (1974) 'Intercropping of Wheat and Sugarcane', *Indian Journal of Agricultural Science*, Vol. 44 (4): 226–30.
Singh, Panjab and N.L. Joshi (1980) 'Intercropping of Pearlmillet in Arid Areas', *Indian Journal of Agricultural Sciences*, Vol. 50 (4), April: 338–41.
Singh, S., R. Singh and O.S. Tomar (1973) 'Intercropping Cotton with Summer Legumes in Punjab', *Cotton Development*, Vol. 2 (4): 9–13.
Singh, S.P. (1981) 'Studies on Spatial Arrangement in Sorghum–Legume Intercropping Systems', *Journal of Agricultural Sciences*, Cambridge, Vol. 97 (3), December.
Singh, S.P., R.C. Gautam and V.R. Anjaneyulu (1979) IARI, New Delhi, 'Competition-free Period to Intercrop Component – A New Concept', mimeograph, prepared for the International Intercropping Workshop, 10–13 January 1979, ICRISAT, Hyderabad, India.
Singh, Surinder and R.C. Singh (1976) 'Economics of Mixed Cropping in Pigeon Pea (Arhar) under Haryana Conditions', *Haryana Agric. Univ. J. Res.*, Vol. 6 (3/4), Sept/Dec: 171–5.
Sinthuprama, Sophon (1979) 'Cassava and Cassava-based Intercrop Systems in Thailand', in Weber and Campbell (eds) (1979).
Sivaraman, G.A. (1973) 'Bisagi Moong, a Profitable Intercrop in Sugarcane', *Farming Facts*, Vol. 7 (5): 10–12, March.
Slikkerveer, L.J. (1994) *Indigenous Agricultural Knowledge Systems in Developing Countries: A Bibliography*, Indigenous Knowledge Systems Research & Development Studies, No. 1, Leiden/Nairobi/Bandung/Chania, 185 pp.
Soleri, D. and D.A. Cleveland (1988) 'Diversity and Small-scale Traditional Agriculture', *New Alchemy Quarterly*, Summer (32): 9–10.
Sooksathan, I. and R.R. Harwood (1976) 'A Comparative Growth Analysis of Intercrop and Monoculture Plantings of Rice and Corn', IRRI Saturday Seminar, 21 February.
Standifer, L.C. and M.N. Bin Ismail (1975) 'A Multiple Cropping System for Vegetable Production Under Subtropical, High Rainfall Conditions', *Journal of American Soc. Hort. Sci.*, Vol. 100 (5): 503–6.
Stern, V.M. (1961) 'Interplanting Alafalfa in Cotton to Control Lygus Bugs and Other Insect Pests', *Tall Timbers Conference on Ecological Animal Control by Habitat Management, Proceedings*, Vol. 1: 55–69.
Stewart, K.A. (1980) 'Intercropping', *MacDonald Journal Quebec*, Vol. 41 (3): 12–13, March.
Su, N.R. (1975) 'Fertility Management in Multiple Cropping Systems', *Soils and Fertilizers in Taiwan*, Vol. 22: 47–62.
Swart, P. de (1986) *Mogelijkheden voor het Gebruik van Indiaanse Kennis van Technieken in Duurzame Landbouwsystemen in het Amazone-Gebied*, Wageningen: Projektgroep Alternative Methoden voor Land- en Tuinbouw, Landbouw Universiteit Wageningen.

Tarhalkar, P.P. and N.G.P. Rao (1979) IARI, Regional Station, Rajendranagar, Hyderabad 30, 'Genotype-Plant Considerations in the Development of an Efficient Intercropping System for Sorghum', mimeograph.

Terhune, E.C. (1976) 'Prospects for Increasing Food Production in Less Developed Countries Through Efficient Energy Utilization', St. Louis: Energy and Agriculture Conference.

Thanee, A.P. (1992) 'Farmer as Scientist: Bringing the Farmer's Knowledge to Research', *Pacific Viewpoint*, 33 (2).

Theunissen, J. and H. Den Ouden (1980) 'Effects of Intercropping with *Spergula arvensis* on Pests of Brussels Sprouts', *Entomologia Experimentalis et Applicata*, Vol. 27 (3).

Thung, M. and J.H. Cock (1979) 'Multiple-cropping Cassava and Field Beans: Status of Present Work at the International Centre of Tropical Agriculture', in Weber and Campbell (eds) (1979).

Tiwana, M.S. and K.P. Puri (1979) 'Intercropping of Napier Bajra Hybrid (NB-21) with Winter Forage Crops', *Indian Journal of Dairy Science*, New Delhi, Indian Dairy Association, Vol. 32 (4), December: 419–23.

Toba, H.H., A.N. Kishaba, G.W. Bohn and H. Hield (1977) 'Protecting Muskmelons against Aphid-borne Viruses', *Phytopathology*, Vol. 67 (11), November: 1418–23.

UNCED (1992) *Agenda 21*. Rio de Janeiro, Brazil.

Varma, M.P. and M.S.S.R. Kanke (1969) 'Selection of Intercrops for Cotton in India', *Experimental Agriculture*, Vol. 5 (3): 223–30.

Veeraswamy, R., R. Rathnaswamy and G.A. Palaniswamy (1974) 'Studies on the Mixed Cropping of Red Gram and Groundnut under Irrigation', *Madras Agricultural Journal*, Vol. 61 (9): 801–2.

Verghese, P.T. (1976) 'Intercropping in Coconut Garden', *Intensive Agriculture*, Vol. 14 (9), November: 11–13.

Waddell, E. (1972) *The Mound Builders*, Seattle: Univ. of Washington Press: 38–61.

Wahua, T.A.T. and D.A. Miller (1978) 'Effect of Intercropping on Soybean $N*d*2$ Fixation and Plant Composition', *Agronomy Journal*, Mar/Apr: 292–5.

Wahua, T.A.T., O. Babalola and M.E. Aken'ova (1981) 'Intercropping Morphologically Different Types of Maize with Cowpeas: LER and Growth Attributes of Associated Cowpeas', *Experimental Agriculture*, Cambridge, Vol. 17 (4), October: 407–13.

Warren, D.M., L.J. Slikkerveer and D.W. Brokensha (eds) (1994) *The Cultural Dimension of Development: Indigenous Knowledge Systems*, IT Studies in Indigenous Knowledge and Development, London: Intermediate Technology Publications, 582 pp.

Weber, E.B.N. and M. Campbell (eds) (1979) *Intercropping with Cassava: Proceedings of an International Workshop held at Trivandrum, India, 27 Nov.–1 Dec. 1978*, Ottawa: IDRC.

Wijesinha, Anila, T. Federer, J.R.P. Carvalho and Thomas de Aquino Portes (1982) 'Some Statistical Analyses for a Maize and Beans Intercropping Experiment', *Crop Science*, Madison, Wis., Vol. 22 (3), May/June: 660–6.

Wilken, Gene C. (1987) *Good Farmers. Traditional Agricultural Resource Management in Mexico and Central America*, University of California, Berkeley.

Willey, R.W. (1979a) 'Intercropping – Its Importance and Research Needs. Part 1. Competition and Yield Advantages', Commonwealth Bureau of Pastures and Field Crops, *Field Crop Abstracts*, Vol. 32 (1), January, ICRISAT, 1–11–256, Begumpet, Hyderabad 5000 16, AP India.

Willey, R.W. (1979b) 'Intercropping – Its Importance and Research Needs. Part 2. Agronomy and Research Approaches', *Field Crop Abstracts*, Vol. 32 (2), February: 73–85.

Willey, R.W. (1979c) 'A Scientific Approach to Intercropping Research', prepared for International Intercropping Workshop, ICRISAT, Hyderabad, India, January.

Willey, R.W. and M.R. Rao (1981) 'A Systematic Design to Examine Effects of Plant Population and Spatial Arrangement in Intercropping, Illustrated by an Experiment on Chickpea/Safflower', *Expl. Agric.*, Vol. 17: 63–73.

Willey, R.W. and M.S. Reddy (1981) 'A Field Technique for Separating above- and below-ground Interactions in Intercropping: An Experiment with Pearl Millet/Groundnut', *Expl. Agric.*, Vol. 17: 257–64.

Williams, C.N. (1975) *The Agronomy of the Major Tropical Crops*, Oxford, Kuala Lumpur.

Wilson, G.F. and M.O. Adeniran (1976) 'Studies of the Intercropping of Cassava with Vegetables', International Institute of Tropical Agriculture, in *Proceedings of the Second Symposium on Intercropping for Semi-Arid Areas*, Morogoro, Tanzania, 10–11 May.

Wilson, P.W. and O. Wyss (1937) 'Mixed Cropping and the Excretion of Nitrogen by Leguminous Plants', *Soil Science Society of America*, pp. 289–97.

Yadav, R.L. (1981) 'Intercropping Pigeonpea to Conserve Fertilizer in Maize and Produce Residual Effects on Sugarcane', *Experimental Agriculture*, Cambridge, Vol. 17 (3), July: 311–15.